Newton 1670
Vesalius 1540
Lavoisier 1780
Alfonso the Wise 1250
St. Augustine 400
Scheele 1770
Einstein 1910
Paracelsus 1525
Galileo 1610
Mendeleyev 1870
Copernicus 1540
Aristotle 350 B.C.
Thales 600
The Assyrian Astronomer 700 B.C.
The Egyptian Architect 2800 B.C.
Avicenna 1000

AN ILLUSTRATED HISTORY OF SCIENCE

By the same author

★

SCIENCE PAST AND PRESENT
THE CENTURY OF SCIENCE
THE WORLD OF SCIENCE
THE ALCHEMISTS
A HISTORY OF INDUSTRIAL CHEMISTRY

AN ILLUSTRATED HISTORY OF

SCIENCE

by

F. SHERWOOD TAYLOR

M.A. B.SC. PH.D.

ILLUSTRATED BY

A. R. THOMSON, R.A.

FREDERICK A. PRAEGER

NEW YORK

BOOKS THAT MATTER

*Published in the United States of America
in 1955 by Frederick A. Praeger, Inc..
Publishers, 150 East 52nd Street
New York 22, N. Y.*

Library of Congress Catalog Card Number: 55-10589

Printed in Great Britain

PREFACE

IN 1952 THE ROYAL INSTITUTION honoured me by an invitation to give their annual Christmas lectures, and suggested as a subject the historic experiments of natural science. The six lectures amounted to a brief history of science, written around the theme of these experiments, which were demonstrated in a fashion as near as possible to that of their originators. When I came to prepare the lectures for the press, I was loath to substitute a dry description for a living demonstration, and came to the conclusion that a realistic illustration could do far more than the written word. It is true that an illustration depicts events at but a single moment of time, but what it loses in action is more than compensated by its power to reveal the background and personalities. I was extremely fortunate in being able to persuade Mr. A. R. Thomson, R.A., to undertake these illustrations: the results have much exceeded my hopes.

Scientists and historians alike look askance at modern pictures of past events, feeling that the author and artist cannot fail to incorporate details for which no authority can be found. But if the reader accepts these pictures, not as authoritative sources, but as a synthesis of what has been transmitted by documents and what the author and artist know about the ways of life in days gone by, he will find in them the means of forming a visual idea of the men and events that brought science to its present position of pre-eminence. Yet in order that the student may not be tantalized, I have provided an appendix indicating some of the sources which we used in devising these windows on the past.

F. SHERWOOD TAYLOR

ACKNOWLEDGMENTS

IT WOULD BE DIFFICULT FOR me to chronicle the many persons and institutions who gave me generous help in the preparation of the material on which the present book is based. Some of them I attempted to thank verbally at the close of the series: the help of others is acknowledged in the captions or appendix. I am especially happy to acknowledge the help of Mr. L. Walden, Senior Experimental Assistant at the Royal Institution. I am also glad to be able to thank the officers of the Science Museum, who gave me very valuable information and criticism. Finally I wish to thank the persons and institutions who assisted me with information bearing on the illustrations, notably Mr. G. H. H. Wheler of Otterden Place in Kent; Professor C. T. R. Wilson; Sir Lawrence Bragg; Professor L. Rosenfeld; The Royal College of Physicians; The National Portrait Gallery and The Tate Gallery.

My thanks are also due to the following for their kind permission to reproduce illustrations contained in this book: Captain Antonio Simoni: Fig. 5. Crown Copyright, The Science Museum: Figs. 12, 16, 17, 31, 32, 39, 44, 45, 48, 52, 54, 56, 58, 60, 61, 62, 64, 66, 72, 74, 96, 102, 104, 106, 107, 116, and 117. Museum of the History of Science, Oxford: Figs. 15 and 115. Blackfriars: Fig. 18. Crown Copyright, The Master of the Armouries, Tower of London: Fig. 33. Crown Copyright, British Museum (Natural History): Fig. 92. Royal Institution: Fig. 113.

CONTENTS

PREFACE v

ACKNOWLEDGMENTS vii

LIST OF ILLUSTRATIONS x

INTRODUCTION: THE PERIODS OF SCIENCE 1

I THE FIRST BEGINNINGS OF SCIENCE 3

II EXPERIMENTAL SCIENCE BEGINS 52

III SCIENCE IN THE EIGHTEENTH CENTURY 81

IV THE REFOUNDING OF THE SCIENCES 105

V THE BEGINNING OF THE HARVEST 124

VI SCIENCE TAKES THE LEAD 150

APPENDICES

I NOTES UPON ILLUSTRATIONS 168

II STEPHEN GRAY 172

INDEX 175

LIST OF ILLUSTRATIONS

FIG.		PAGE
1.	The periods of Science	1
2.	Egyptian Hour-watchers	4
3.	Egyptian Constellations	5
4.	Telling the time in Egypt	7
5.	Fifteenth century alarm-clock in use	8
6.	John Knox and the sand-glass at St. Andrews	9
7.	Assyrians transporting a statue	10–11
8.	Egyptian metallurgists	13
9.	Thales in the Nile Delta	15
10.	The four qualities and elements	16
11.	The five regular figures	17
12.	The world before Copernicus	18
13.	Simon Stevin shows that different weights fall in the same time	21
14.	Aristotle's biological honeymoon	23
15.	A fourteenth-century astrolabe	24–25
16.	Use of an astrolabe to measure heights	26
17.	The track of the planet Mars	27
18.	The planetary system of Eudoxus	28
19.	The planetary system of Ptolemy	29
20.	A scaph	30
21.	How Eratosthenes measured the earth	31
22.	Pythagoras shows the relation between the pitch and length of strings	33
23.	An Archimedean screw	34
24.	Philo of Byzantium investigates air	35
25.	Hero of Alexandria mystifies the devout	37
26.	Hero's reaction-turbine, the first steam-engine	38
27.	St. Augustine is surprised by the loadstone	39
28.	A primitive compass	41
29.	Maria the Jewess in her alchemical laboratory	42
30.	The Alembic	43
31.	A still with a fractionating column (1512)	44
32.	The first professional chemists assaying gold	45
33.	Incendiary bombs in the Middle Ages	46
34.	Roger Bacon and the cracker	47
35.	A mediaeval pharmacy	49
36.	A Greek clinic	50
37.	The skeleton contemplates mortality	53
38.	Galileo times a ball rolling down a slope	55
39.	Leonardo's drawing of a shower of projectiles	56
40.	Walter Ryff's idea of the trajectory of a cannon ball	57
41.	Galileo's diagram of the parabolic trajectory of a projectile	57
42.	Galileo tries to measure the force of the vacuum	58
43.	Torricelli demonstrates the Vacuum	59
44.	Otto von Guericke tries to evacuate vessels	60
45.	The Magdeburg hemispheres	61
46.	Galileo discovers the constancy of the time of swing of a pendulum	63
47.	Galileo observes the moon	64
48.	Galileo's and a modern picture of the moon	65
49.	Galileo's diagram of the solar system	66
50.	The elliptical orbit of a planet, illustrating Kepler's laws	67
51.	Robert Hooke prepares a flea for the microscope	69
52.	Hooke's engraving of a flea, much reduced	70

FIG. PAGE

53. Antony van Leeuwenhoek and his microscope 71
54. Leonardo da Vinci's drawing of the valves of the heart 72
55. William Harvey prepares for a dissection 73
56. Refraction and reflection as studied in the sixteenth century 74
57. Newton splits up sunlight into the spectral colours 75
58. Newton's rings 76
59. Newton with his original reflecting telescope 77
60. An instrument-maker's trade-card 79
61. Swammerdam's dissection of a may-fly 80
62. William Gilbert's terrella 82
63. Fellows of the Royal Society studying the loadstone 83
64. William Gilbert's electroscope 84
65. Otto von Guericke's electrified sulphur globe 85
66. Hauksbee's electric machine 86
67. Stephen Gray demonstrates electrical conduction 87
68. Du Fay is led to postulate two kinds of electricity 89
69. The Abbé Nollet receives a shock 90
70. The Abbé Nollet shows that the conduction of electricity seems to be instantaneous 91
71. Benjamin Franklin draws electricity from a thunder-cloud 92
72. Coulomb's torsion-balance 93
73. Robert Hooke and Robert Boyle investigate respiration 94
74. John Mayow's apparatus for studying respiration and combustion 96
75. A seventeenth-century pharmacist calcines antimony 97
76. Stephen Hales's apparatus for measuring air 98
77. Henry Cavendish in his laboratory 99
78. Scheele prepares oxygen 101
79. Cavendish's metal eudiometer 102
80. Lavoisier and his wife in their laboratory 103
81. John Dalton's and modern formulae 106
82. John Dalton teaching 107
83. An atmospheric steam-engine at a colliery 109
84. Count Rumford shows that heat is a form of motion 110
85. How two rays of light can produce darkness 111
86. Longitudinal and transverse waves 112
87. Luigi Galvani produces electricity from two metals and a frog's legs 114
88. Alessandro Volta and his batteries 116
89. Sir Humphry Davy demonstrates the arc-light 117
90. Faraday's electromagnet 118
91. H. C. Oersted experiments on the magnetic effect of the electric current 119
92. Scheuchzer's supposed relic of the Biblical Deluge 120
93. Cuvier's illustration of a fossil pterodactyl 121
94. Lines of force of a magnet 124
95. Faraday generates a current from a magnet and a coil of wire 125
96. Edison's dynamo 127
97. J. P. Joule experimenting on the mechanical equivalent of heat during his honeymoon 129
98. Faraday's sketch of the way he rotated a beam of polarised light by means of magnetism 130
99. Sir William Herschel demonstrates invisible rays 131
100. Diagram to illustrate the nature of an electromagnetic wave 132
101. Hertz generates and detects radio-waves 133
102. The spectrum of the star Mira Ceti compared with that of titanium oxide 134

FIG.		PAGE
103.	Crookes demonstrates the deflection of cathode rays by a magnet	135
104.	Sir John Herschel and his photograph (1839) of the 40 foot telescope	136
105.	Robert Bunsen and his spectroscope	137
106.	The nebula in Orion as drawn by Sir John Herschel	138
107.	The same nebula photographed by a modern telescope	139
108.	Faraday liquefies chlorine	141
109.	James Dewar liquefies air	143
110.	Some of Louis Pasteur's flasks	145
111.	Pasteur lectures on the germ-theory of putrefaction	147
112.	The Descent of Man	149
113.	An early model of the structure of zinc blende, made by Sir William Bragg	154
114.	Sir William Bragg and his X-ray spectrograph	155
115.	Distribution of electrons in penicillin molecule, leading to its formula	156
116.	Model of the aluminium atom, as proposed by Niels Bohr	157
117.	Photographs of trails made in the Wilson cloud-chamber	160
118.	C. T. R. Wilson and his cloud-chamber (1911–12)	161
119.	Wollaston's laboratory comes in on a tea-tray	164
120.	The modern physicist's equipment	165

INTRODUCTION

THE PERIODS OF SCIENCE

THE ROOTS OF SCIENCE PASS down beyond the deepest strata of history into the darkness which surrounds the beginning of man himself. In the history of man's dealings with matter, we mark four stages. First, an enormous stretch of time before the beginning of history in which the men who worked upon things necessarily came to understand their ways and invented practical rules of craft, rules which were transmitted by the word and example of those who had learned to do such tasks as fire-making, flint-knapping, spinning or pottery. Such crafts imply observation of matter and knowledge about it, and thus far resemble a science, but until the age of writing, the volume of knowledge

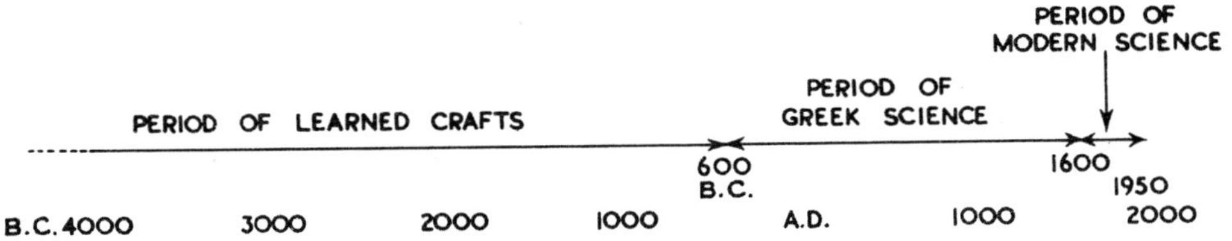

Fig. 1. The Periods of Science.

handed on could be but small, and its tenure was as precarious as the life of man.

In the years between 4000 and 3000 B.C. some communities organized themselves into larger units. Civilization began and with it, in Mesopotamia and Egypt, the art of writing. The useful arts, such as building, smelting, healing and time-telling, were far more highly developed than ever before, and the rules of these arts were recorded in writing. Man entered on the second period of his mastery over matter, that in which the knowledge and rules that were the fruit of his observations could first be accumulated and preserved.

No comparable advance was made until, some three thousand years later, the Greeks, not content with recording practical arts, began to look at the world as a whole and tried to see a pattern in it, to discover reasons why it was such as they thought it to be. Thus, about 600 B.C., began theoretical science, intended to fulfil the desire to know rather than the desire to achieve. The Greeks reasoned better than they observed, so that, too often, they based beautiful theories on unsound data. Those theories took the whole world captive, and for twenty-two centuries, Western science was Greek science;

for the Romans, the Arabs and the men of mediaeval Europe did no more than enlarge on Greek ideas.

Yet the science handed on by the Greeks did not bear the test of time. Gradually it became apparent that in many respects it did not assert or predict what men observed. The fifteenth, sixteenth and seventeenth centuries saw its downfall; and in its place the establishment of a new science based on observations of proven reliability and logical reasonings about them. The scientists of the three centuries from 1650 to the present day have been occupied in ascertaining true facts, discovering the rules that connect them, and forming theories that account for those rules. The results are to be seen in comparing the world of 1650 with that of 1950. The effectiveness of the kind of science we practice today has made men believe that they have found the best way of regarding the world around them: yet in other times they also have thought this and been proved wrong.

CHAPTER ONE

THE FIRST BEGINNINGS OF SCIENCE

IT IS REASONABLE TO ASSUME that the man of seven thousand years ago had hands and eyes and brain as good as those of the man of today, but we know very little of what he did with them. Some of the people who lived before civilization began could make fire, shape flint, weave baskets, make pots, paint and carve, build boats and practice the skilled arts of hunting and agriculture; but their work, skilled and realistic as it was, lacked the greatest property of science, namely that it is recorded, so that what one generation does and learns, the next generation can know and use. Accordingly we may say that something recognizable as kin to science began in Egypt and Mesopotamia five or six thousand years ago, when men first recorded in writing some general rules about things. That meant that there were men who understood the sciences and who were able to write. These were the priests, the men who understood the difficult things, such as writing, the ways of the stars, the arts of architecture, metallurgy and medicine, the secrets of the gods and whatsoever else was thought to be wisdom.

The loftiest of the beings that man can observe are those in the heavens. The study of them, which we call astronomy, goes back to the years before 3000 B.C., centuries before the Great Pyramid was built.

Why were these ancient people concerned to know about the heavenly bodies? To the man of ancient times the sun and moon, planets and stars, seemed to cause and rule the growth of the crops on which life depended; the fertilizing floods of the Nile, the great event of the Egyptian year, was seen to coincide with the time that certain stars first began to be seen in the morning sky. The sun and the lesser lights seemed to them to exercise a power, which ranked them as gods, prime causes or at least intelligent agents: to study the heavens was, therefore, a matter both of religion and of practical use. It is believed that the stars were first grouped into constellations between 4000 and 3000 B.C. by the Sumerians, inhabitants of Mesopotamia; between 3000 and 2000 B.C. the Egyptian priest-astronomers learned to make an annual calendar by their aid and also learned to tell the time by them.

For every period of science, there are two important questions to be asked: 'What did men want to know?' and, 'What means had they of discovering it?' Accordingly we must ask how the Egyptians observed the heavenly bodies. We know that there were 'hour-watchers' who sat on the roofs of the temples and noted the position of the stars. These Egyptian astronomers

Fig. 2. Egyptian hour-watchers observing the stars from the temple roof.

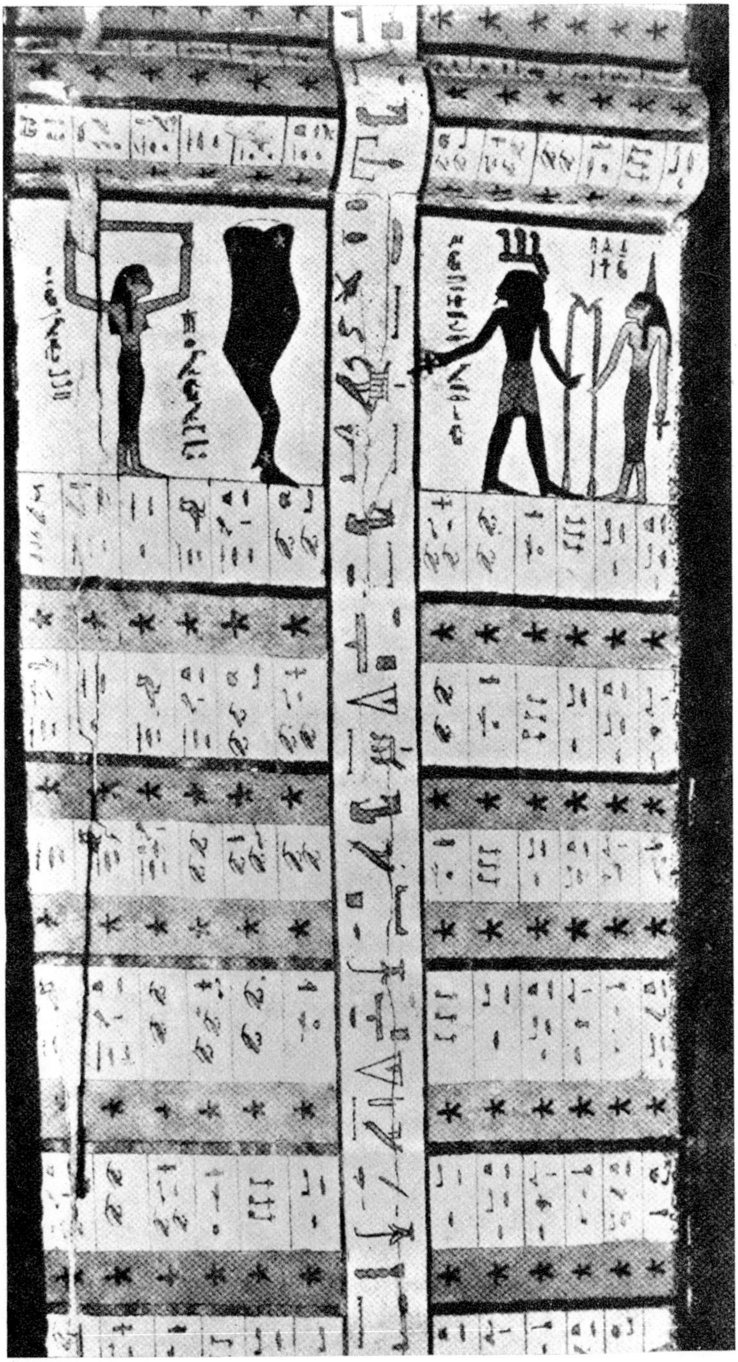

Fig. 3. Constellations pictured on the lid of an Egyptian sarcophagus. The Great Bear appears as the Bull's Haunch.

had no telescopes, of course, and perhaps not even the means of measuring angles, but they are known to have employed the instrument shown in Fig. 2. It was the rib of a palm-leaf, having a slot cut in the base and a plumb-line hanging from the tip. It seems likely that, by the aid of these, two hour-watchers could set themselves in line with the pole-star, low on the horizon. They could then describe the position of stars relative to the watchers: thus a conspicuous star could be described as 'over the hour-watcher's right shoulder', another as 'above the hour-watcher's head'. It was thus easy to see that the stars circled daily about the poles and to tell the time by them. But in fact the stars do not keep time with the sun, but rise about four minutes earlier every day. Thus, as the year goes on, a succession of new stars become visible before the light of the rising sun overpowers them, and by the appearance of these stars the progress of the seasons could be reckoned.

Fig. 3 shows part of an Egyptian star-calendar, intended to record the stars that just become visible, in each ten-day week or decan, before the sun rises.

The motions of the sun were used to tell the time of day by shadows cast on sundials. Doubtless the first dial was no more than a stick planted in the ground. The earliest surviving instrument specially made to tell the time by the sun may date from 1500 B.C., but that does not mean that such things were not used much earlier. In our latitudes the sun is never very high in the sky, and in winter indeed it does little more than crawl around the horizon; so our sundials are made to measure the way the shadow travels round a horizontal dial. But in parts of Egypt the sun in summertime rises in the east and passes overhead in an east-west line to set in the west. Accordingly Egyptian dials mark the length of the shadow rather than its direction. The shadow was cast along a graduated bar, and the graduation of it argues some little knowledge of geometry. To learn to divide space, by measuring rods, was an achievement, but to divide time was indeed a greater discovery.

The Egyptians also invented the water-clock, in which time is measured by the volume of water running out of a vessel pierced with a hole. The hours were read off by observing the level of the water in the vessel. The accuracy of such a clock depends on the water falling equal distances in equal times. As the depth of water becomes less, so the pressure falls and the water runs more slowly, but in the clock illustrated the effect is compensated by making the vessel conical so that, as the level falls, the escape of less and less water is needed to lower the surface by a given distance. Water-clocks could never be accurate, however, for the viscosity of water varies with its temperature, and so on a hot day the clock would take less time to mark an hour than on a cold day.

So the Egyptians and Babylonians learned to measure time by the sun and stars and by the flow of water and after that nobody introduced any important new principle into time-telling for some three thousand years, after which vast lapse of time, about A.D. 1250, some unknown person invented a mechanical clock. A mechanical clock must have its rate controlled by something that oscillates at an unvarying rate. In a mediaeval clock this was a foliot, a heavy horizontal weighted rod or ring, which (like the modern pendulum) allowed the escapement wheel to move forward by one tooth each time it oscillated. But the time of each swing of the foliot remained the same only as long as the force behind it did not vary and the force depended on the pull of the weights and the friction of the mechanism. If it kept time to twenty minutes a day it did well; it was normally set by a sundial and if the sun was invisible for a few days, the mediaeval clock was soon well and truly lost.

Fig. 4. Telling the time by sun-dial and water-clock in ancient Egypt.

Fig. 5. An alarm-clock in use in a fifteenth-century Italian monastery. These were among the earliest portable mechanical clocks. The ring at the top of the clock is the foliot, the oscillation of which keeps its rate constant.

Fig. 6. The sand-glass was commonly used to time sermons and many examples of sand-glasses or their holders are found in British churches. Such a holder is attached to the pulpit at St. Andrews, traditionally associated with John Knox, here pictured in one of his more eloquent moments.

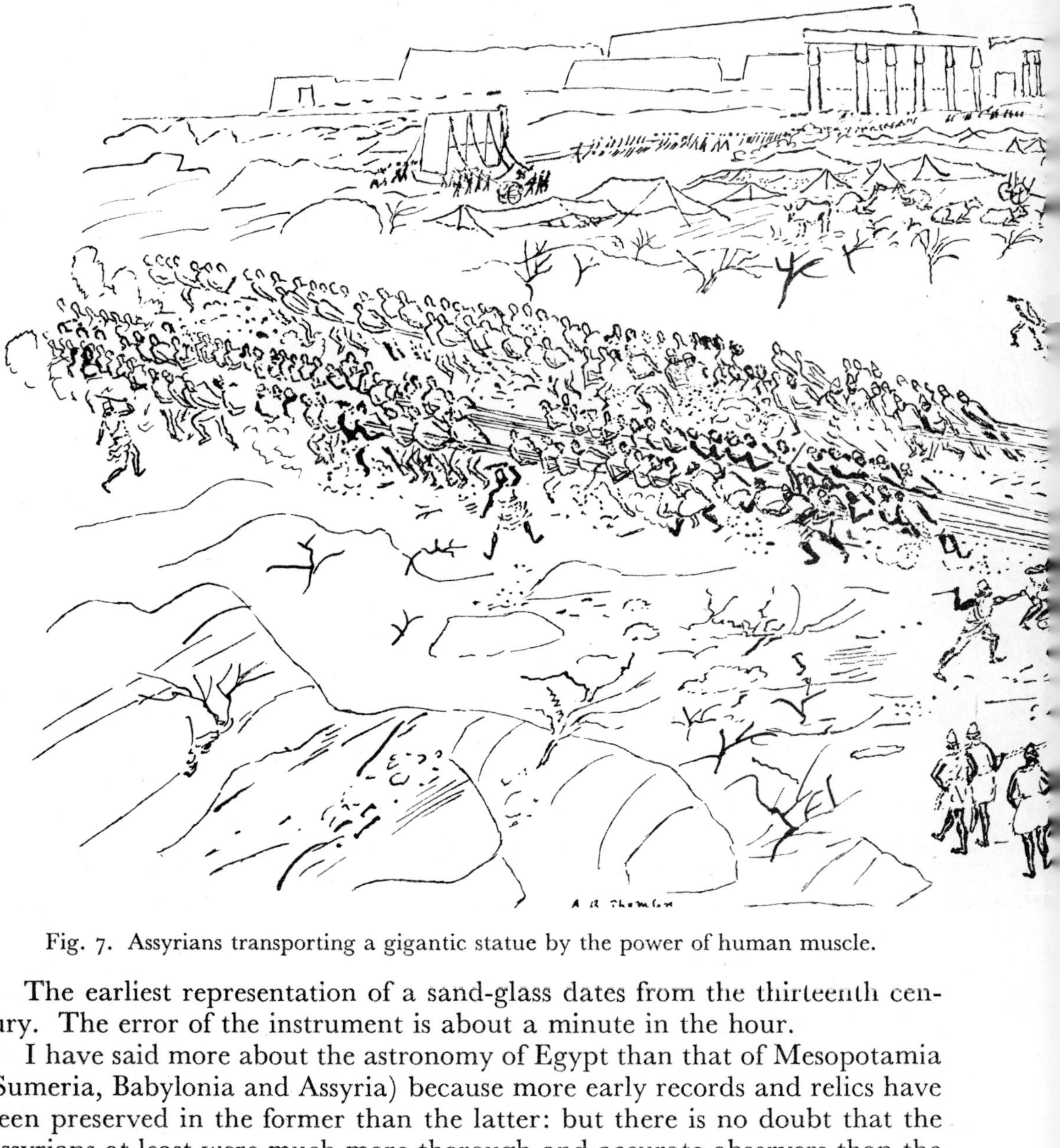

Fig. 7. Assyrians transporting a gigantic statue by the power of human muscle.

The earliest representation of a sand-glass dates from the thirteenth century. The error of the instrument is about a minute in the hour.

I have said more about the astronomy of Egypt than that of Mesopotamia (Sumeria, Babylonia and Assyria) because more early records and relics have been preserved in the former than the latter: but there is no doubt that the Assyrians at least were much more thorough and accurate observers than the Egyptians and had at their disposal a much more elaborate arithmetic and even algebra. They named the constellations and invented the signs of the Zodiac, and from them seems to have originated the enormously influential doctrines of astrology.

We know that the Egyptians and Assyrians were able to weigh accurately and had invented balances and sets of weights good enough for goldsmith's work: so we arrive at the cardinal fact that these early peoples had learned to determine with fair accuracy the three fundamental quantities of science—mass, length and time. Their observation of the stars and invention of a calendar amount to a proof that they comprehended the prime axiom of science, namely *the regularity of nature*. They learned that the heavenly bodies went through regular cycles of movement that recurred in precise periods—for example, that eighteen years and ten days after an eclipse they might expect another eclipse, that certain stars would become visible in a regular order as the year progressed. This accumulation of *exact recorded rules* capable of predicting what would happen to things was of the essence of science.

The men of the first civilizations were great builders; so they had to become civil engineers and learn the way to measure and lay out land, to plan buildings, to cut stones to precise angles so as to make inclines of a given slope. The Great Pyramid, dated as *c.* 2800 B.C., is a miracle of precision, its angles and sides having been laid out, it would seem, almost with the accuracy of modern science. It has been much damaged, but it would seem that the angles of the corners differed from 90 degrees by only $\frac{1}{300}$ degree and the sides of 254 yards were equal within $\frac{2}{3}$ inch; or so they say. This kind of feat the Egyptians did by a simple practical geometry—and a vast expenditure of skill, time and labour to put the geometer's plans into practice. They did not, however, invent a formal geometry with proofs and propositions—that was a Greek idea—but they seem to have discovered the properties of figures and angles in an experimental sort of way: their geometry, in fact, was more like science than mathematics.

We are on less certain ground when we talk about the chemical inventions of the ancient peoples, chief of which was the art of smelting metals. The Egyptians and Assyrians were the first metallurgists and very accomplished ones. It is true that their methods were probably very simple and amounted to little more than burning the ores with charcoal. But even this involves recognizing the ores, mining them, separating the gangue or rocky parts from the mineral, reducing them to the right-sized fragments, and heating them with enough draught or blast to keep up the heat but not enough to re-oxidize the metal. To discover the smelter's technique for the six metals gold, silver, copper, tin, lead, and later iron, was to form a great body of practical knowledge and we suppose this to have been recorded, since metallurgy was a temple-industry.

The first metal that the Egyptians used was gold, which is found as native gold and does not have to be smelted; the second seems to have been copper. We do not know how the copper ores were found to contain a metal. At a very early date the Egyptians used the conspicuous mineral malachite for making a green eyepaint: we might guess that some lady in a raging temper threw her cosmetic box into the fire, and that later someone raked out of it a bit of heavy red metal—something rather like gold, already known and esteemed. Be this as it may, all that has to be done to malachite to make it into copper is to burn it in a bright charcoal fire. Once it was known that this heavy mineral could be made into copper, it would be natural to try other heavy minerals, and thus it was, perhaps, that the other metals were discovered.

The Egyptians and Mesopotamians knew how to make glasses and glazes for pottery, recipes for which they recorded in great numbers. They were

Fig. 8. An Egyptian metallurgical temple-workshop. Gold is being fused in fires urged by bellows or the blow-pipe—a clay-tipped reed. In the background is a gold-beater, and an artisan, with a scribe, weighing gold rings.

skilled, in fact, in chemical arts, but as far as we know they had no idea of chemistry, no theories about how or why these recipes worked. But this work certainly taught them that matter could be transformed in strange ways, that dull stones produced shining metal, or brilliant glass, and that these were produced, not in any capricious way, but by rule and law: and that is the root-idea of chemistry.

Finally these early peoples studied medicine. They invented some quite good methods of treatment and had a large armoury of drugs. Most of these were useless, but as there are physicians today who say the same of a few of the drugs in our official pharmacopoeia, we shall not expect too much of the men of four or five thousand years ago. The Egyptians had rather elementary ideas about that difficult subject, anatomy, but they must at least have come to realize that man's body is no less ordered in its construction than the rest of the world.

Until perhaps 700 B.C. science dwelt in Egypt and Mesopotamia and we do not know that the achievement of these nations was paralleled anywhere else, though the Far East may have accomplished much of which no record now remains. The Chinese certainly were great craftsmen as early as 1500 B.C. but we know very little about their achievements in the sciences: the same is true of ancient India, where we know there were civilizations and high craftsmanship as early as in Egypt and Mesopotamia; but the Indian records are scanty and cannot be certainly dated till very much later.

About 600 B.C. the people of Asia Minor and, soon after, of Greece found a new kind of interest in the world. Till that time such studies as mathematics, astronomy or medicine had been pursued because they were useful, a superior sort of learned craftsmanship; but the Greeks, who were becoming a settled prosperous people with leisure to think, began to study science for its interest rather than its use. They began to ask great far-reaching questions. What are things made of? Where do they come from? What is change and motion? Can things really change and move and, if so, how? What is the universe? Is it finite or infinite? Had it a beginning and will it have an end? What is life and soul?—and so forth. These questions are on the borderline between philosophy and science—a line that the Greeks were not much concerned to draw. They can be discussed as theology; men may say, 'Things are as they are because God made them so', and while this is true, and seems to have been enough for the very ancient civilizations, it is not all that the Greeks wanted to know. As Robert Boyle said, he would be a dull fellow who, inquiring about a watch, was satisfied with the answer that it was made by a watchmaker. So the Greeks tried to answer the great questions by observation and reasoning, that is by science, but they did not realize how difficult a task they had set themselves; some of these questions science has not yet answered, and some may be unanswerable.

Yet the world needed to ask these questions of science; and, to give the Greeks their due, they thought of most of the answers, right and wrong, that could be given. The Greeks had as good heads as any people that has been, and used them as well: but science needs not only the head, but also the hand and the eye, and only very late in their career did the Greeks realize that thinking was not enough. So most of their science was based on the ordinary commonsense view of the world, unaided by any specially accurate kind of observation; on this unsound basis they erected brilliant theories, theories of how the world was made and why it behaved as it did.

Fig. 9. Thales, the Ionian philosopher, is here imagined as on his travels in the Nile delta, and seeing there the apparent conversion of water into earth.

The Greeks' first problem was the question, 'What are things made of and where do they come from?' For some reason man has already thought that the myriads of different things and stuffs of which the world is composed are all made up of a very few simpler materials combined in different ways. Modern science confirms this ancient faith. The Greeks tried to discover the elements of the world, the stuff or stuffs that was the origin of everything. The first Greek philosopher, Thales, called this primal stuff 'water'. Why did he think that everything could be made from water? He has not told us; but I suppose that, being a travelled man, he had seen the Nile laying down the earth of the delta and covering the fields with fertilizing mud: earth was being made from water. The sky, too—did not rain and dew fall from it? Did not water left in an open dish turn into air? Life itself could not continue without water—surely this was the source of every being? Yet this was not the only possibility, for a later Greek guessed that everything took its origin from

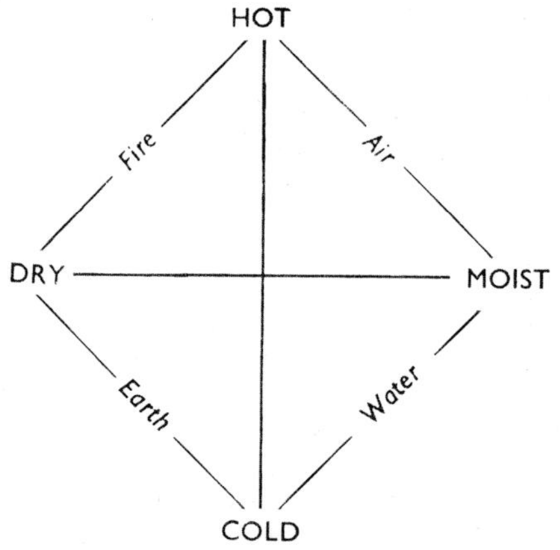

Fig. 10. The four qualities and the elements that shared them

air. Air would grow thick and misty and turn to water, and water would lay down earth. Moreover, air was breath—breath of life and something like soul. A third Greek had all things to be fire, the living moving element, stuff of the sun and heavenly bodies—a conception rather like that of energy today. Other philosophers were not content with one element, but would have three or four. So it went on, but after many theories had been propounded, a very great philosopher and teacher, Aristotle, tried to make a critical selection from all these theories and compile a sort of encyclopaedia of the best-accepted knowledge of his time. His writings were preserved while those of most of the other Greek scientists perished; so it was his ideas that were passed on to the men of later times and came to be accepted, almost as articles of faith. So let us now see what Aristotle thought about the constitution of things.

Fig. 11. The five regular figures. *Top line:* Icosahedron (water); Dodecahedron (the celestial element). *Below:* Tetrahedron (fire); Cube (earth); Octahedron (air).

First of all, he supposed that all the properties of things could be accounted for if they were supposed to be hot or cold or moist or dry in varying degrees. The variation of these qualities he explained by supposing there to be four elements, having the four possible permutations of those qualities. Fire was hot and dry, air was hot and moist, water was cold and moist, earth was cold and dry. Everything on earth was made of combinations of these elements, and all changes (other than motion) depended on alterations of the proportions of these. Certainly he and his followers could explain a great many changes in that way: but there was not much visible evidence for the very existence of his earth, air, fire and water in things. Aristotle scarcely attempted to produce such evidence. His later followers said, 'Put a log in the fire: you will see *water* ooze out of the end; smoke will issue, which is a sort of *air*; flames will appear, which are *fire*: and ashes will be left, which are *earth*. So there are earth, air, fire and water in a log.' But they

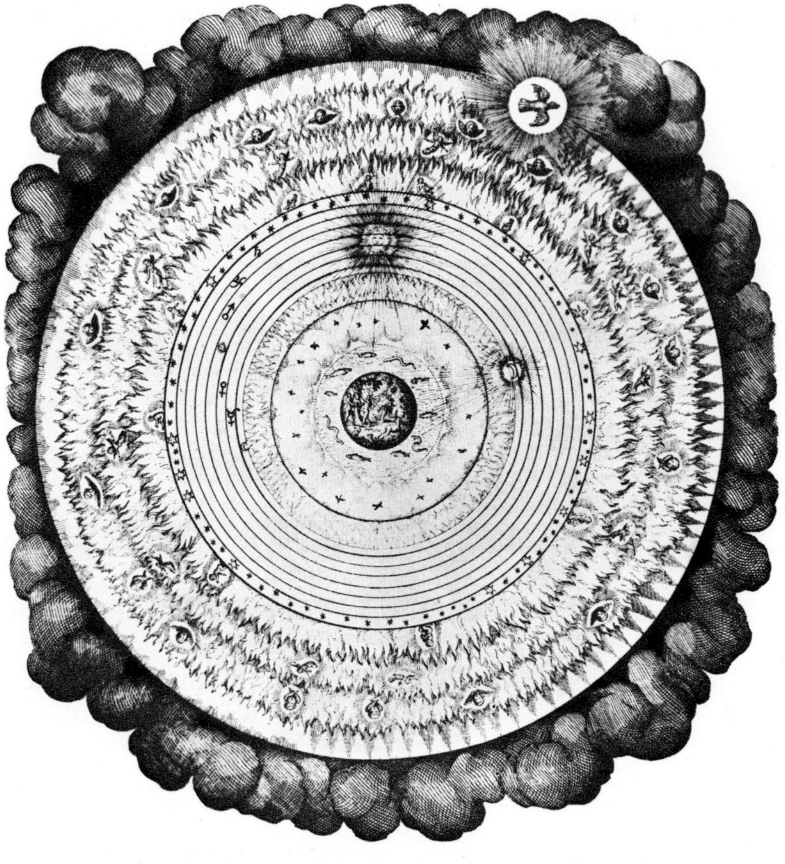

Fig. 12. The world before the time of Copernicus and Galileo. The universe is a sphere. From the centre outward, we see the terrestrial region with successive layers of earth, water, air and fire, ending at the sphere of the moon. Beyond this are the spheres of the planets and Sun, then that of the fixed stars. Outermost of all is the Empyrean sphere, habitation of God and the saints. This figure was drawn while Galileo was still alive.

could not apply this to, let us say, gold or glass. All the same the four-element theory had its uses, at least as a way of describing things. Aristotle was so universal a genius that it was hard to believe him to be wrong, and so the four-element theory became firmly fixed in the heads of all scientists up to the sixteenth century, and was not really dismissed until the end of the eighteenth.

But what about the sun, moon and stars? Were they made of any of these elements? By no means: the four earthly elements changed one into another, but the sun, moon and stars were imperishable. They moved but they did not seem to come to be or pass away. They must therefore be of some other element.

Plato, who was Aristotle's teacher, had been much impressed by the fact that the Greek geometers had discovered that there could be only five regular solid figures with all their sides made up of identical regular polygons and without re-entrant angles. He suggested that they were the materials of the elements. Fire consisted of tetrahedra—small and sharp and spiky; earth of cubes—very solid; water of icosahedra, rounded and slippery; air of octahedra (—but why?); and the material of the heavens of dodecahedra, which could not be made out of the triangles on which the other figures were based. After that, the idea of five elements was generally accepted, four for the earth, one for the heavens.

Now Aristotle, having satisfied himself that these five elements, earth, water, air, fire, and the fifth element, were the basis of things, went on to ask why and how things move, both in the heavens above and on the earth below. And he gave as a reason that *each element goes naturally to its proper place*, unless it is forced to go somewhere else. Fire went naturally up—away from the centre of the earth (for the Greeks and everyone after them knew the earth was a sphere), till it reached the surface of the sphere of the moon where the celestial world began. Earth sank down as near the centre as it could reach. Water sank below air, but above earth. Air rose above water but stayed below fire. Fig. 12 is a mediaeval diagram of the universe. Starting from the centre, earth, water, air, fire and the heavenly regions, are seen all in their proper places.

The latest followers of Aristotle used to illustrate the sorting out of the four elements into their places by shaking up a bottle with four liquids that did not mix. Mercury (or sometimes ashes) would represent earth, saturated potassium carbonate solution (oil of tartar) represented water, alcohol air, and spirit of turpentine fire. No matter how shaken up, they sorted themselves out again into four layers, in the same order; and so should earth, water, air and fire, if they were left alone.

In the heavens the fifth element had a natural circular motion. Clearly the stars moved in circles and so Aristotle adopted the idea that everything in the sky went round and round in circles because to do so was part of the nature of the element they were made of. The circle was the perfect figure and obviously right for such perfect beings as stars and planets. Everyone was so convinced of this that Kepler in 1609 found it hard to persuade himself of the truth of his own discovery that a planet could move in an ellipse.

Then Aristotle asked himself what laws expressed the motion of ordinary heavy bodies, here on earth; and he came to wrong conclusions that led science on a false trail for some eighteen centuries. I suppose that he watched heavy bodies like stones and light bodies like leaves falling through air or sinking through water. He came to the natural conclusion that the heavier a body was, the faster it fell, and the thicker the stuff it fell through, the slower it fell. A stone fell quicker than a leaf, and it sank more slowly through honey than through water. In a general sort of way he was right, but if you are going to found a science on observations they ought to be exact. Aristotle did not make any measurements nor did he grasp the idea of the acceleration of falling bodies; and he came to the lamentable conclusions:

(1) That the speed of a falling body is proportional to its weight, e.g. a 10-pound stone falls ten times as fast as a 1-pound stone;
(2) That the speed of fall is inversely proportional to the resistance of the medium through which the body falls.

Well, this simply was not true. But worse was to come. We have seen that he believed that the speed of fall of a body is inversely proportional to the resistance of the medium it fell through. A vacuum would have no resistance: therefore a body would fall at infinite speed through it. Which, Aristotle said, is absurd, so it is impossible that a vacuum can exist. But if there cannot be a vacuum, there cannot be atoms, for these must move and so, if they existed, there would have to be vacuum between them.

What a tangle resulted from these clever arguments from untrue facts! Could there be a better illustration of the danger of jumping to conclusions without testing or even stating the observations on which they were based? And did nobody else test Aristotle's conclusions about falling bodies? As far as we know, nobody whatever did so until, some time before 1586, the Dutch mathematician and economist, Simon Stevin, performed the experiment and put the answer in a book written in Dutch, which no one but Dutchmen could read and so was largely neglected until Galileo had announced the same result.

> 'The experiment against Aristotle is this: let us take two leaden balls, one ten times greater in weight than the other, which allow to fall together from a height of thirty feet upon a board or something from which a sound is clearly given out, and it shall appear that the lightest does not take ten times longer to fall than the heaviest, but that they fall so equally upon the board that both noises appear as a single sensation of sound. The same, in fact, also occurs with two bodies of equal size but in tenfold ratio of weight.'

The crack of those balls on the board ought to have finished Aristotle's physics, but human beings are very difficult to unconvince, and it took fifty years of hard work and harder words to do it. Aristotle's ideas about bodies undergoing 'violent' motions, e.g. projectiles, were equally astray: we shall hear more of these in the next chapter.

Fig. 13. Simon Stevin, before 1586, disproved Aristotle's theory of falling bodies, by dropping on to a board two different weights which arrived so near together as to make but one sound.

Aristotle was out of his element in physics, but he was the best biologist between the beginning of time and the seventeenth century. Indeed he may be said to have founded biology, and all his observations about animals, especially sea-beasts, are first rate. Indeed some of his observations which later naturalists thought to be absurd have in recent years been found to be perfectly correct. He is the first person whom we know to have made dissections, whereby he discovered something of the internal differences of animals, though he could discover very little about their mysterious inner workings. He listed and compared the kinds of animals that he knew, and classified them in a sensible way, choosing significant and important characters by which to distinguish his classes. His was the first idea of a ladder of nature, a scale of beings increasing in complexity; but, unlike the biologists of today, he did not regard it as an evolutionary ladder, for he supposed species to be quite unchangeable. What Aristotle saw in the living beings around him was purpose and absence of haphazard; all these creatures seemed to him, as indeed they are, wonderfully adapted to the end to which they strive. Aristotle was not only a great biologist, but also a great teacher. He seems to have gathered round him a school of biologists, and his pupil Theophrastus has left us a book on botany, which shows the same careful observation as his master's. After the time of Aristotle and Theophrastus, we find some capable writers on medicine, but none worthy of the name of a biologist for more than fifteen hundred years. The books written on animals during that time were feeble credulous productions and the books on plants were mere herbals. It was only in the thirteenth century that a true biologist appeared, namely Albertus Magnus, who took up Aristotle's work and enlarged it by original observations and criticisms. So despite his errors in physics, Aristotle was a great scientist, and he was perhaps the greatest of all philosophers—logician, moralist, art-critic, political theorist. The world had to prove him wrong on some matters, but there has never again been so truly great a man.

The prime work of science seems to us today to be observation and experiment, in the laboratory and in the observatory, and it seems that as time went on the ancients became more attracted to this kind of work. As we have seen there had been a tradition of astronomical observation for thousands of years before the Greeks, and in their time this was being carried on by them and by the Chaldeans. We do not know much about the Greeks' observations. They were evidently pretty bad, since all through their period of work they took the angle subtended by the moon to be 2 degrees instead of 30 minutes—just four times what it ought to be. To put it more simply, the moon is in fact covered by a silver threepenny bit held at arm's length, while the Greeks for the best part of a thousand years thought it was of the size that would be covered by a half-crown.

Fig. 14. Aristotle examines an octopus on the beach of Mytilene (Lesbos), watched by his bride, Pythias.

Fig. 15. The two sides of an English fourteenth-century astrolabe.
Right: The sights R, which read off the elevation of the body observed on the scale S.
Left: M, the movable star-map. L, a pointer indicating a star. E, the circle showing the position of the sun among the stars on each day of the year. P, the plate with its scale of altitudes. A, the rule which indicates the time.

But there is one instrument, invented by a Greek astronomer, though we do not know by whom, which remained in use from their time right up to the seventeenth century. This is the *astrolabe* and it is very typical of the Greeks who were poor observers but very good mathematicians. On one side is a

graduated circle with sights for observing the altitude of the sun or the stars with very moderate accuracy—which is all the Greeks seem to have asked: the other is a highly ingenious geometrical device for finding the time from the height of the sun or a star. To measure the angle a star makes with the horizon, one hangs or holds the astrolabe from its ring and adjusts the sights till one sees the star through both of them: the observer then reads off the angle—but only to an accuracy of about one degree instead of one second, as today. Suppose the star is Sirius and the altitude in 20 degrees. Then, on the other side of the astrolabe he finds Sirius on the movable star-map. He then turns this till Sirius lies on the 20 degrees circle on the plate below. Next he looks on the star-map at the eccentric circle which shows the place of the sun among the stars for each day of the year and he finds the place for the day of observation. The sun appears to go round the earth once in twenty-four hours, so its place in the sky acts as a 24-hour clock and gives the time. So by putting the pointer over the place of the sun the time is shown on the outer edge. Thus the astrolabe works as a dial for day or night as well as an instrument for observation. Its disadvantage is that it is graduated for one particular latitude and, for any other, another plate must be inserted.

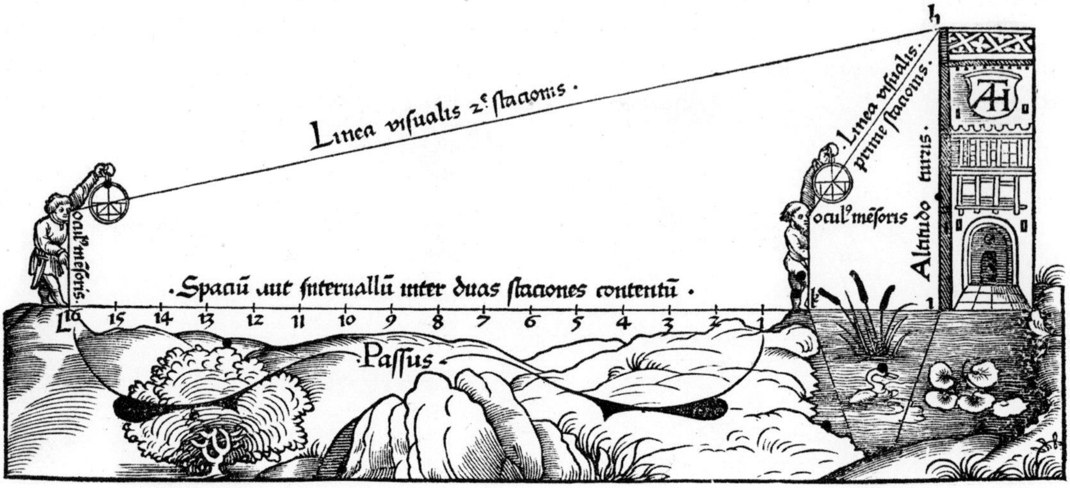

Fig. 16. The use of astrolabes to measure the height of a tower.

The astrolabe could be used as a navigating instrument to find the latitude, as a sextant is used today. It could also be used for surveying, either horizontally where we now use a theodolite or vertically, as we use a level. Fig. 16 shows its use soon after A.D. 1500 for measuring the height of a tower.

The men of the Middle Ages invented numerous astronomical instruments, but there is no reason to suppose them to have been appreciably more accurate than those of the Greeks: the first serious attempt towards precision was made by the Dane, Tycho Brahe, in the years between 1570 and 1600.

The reputation of the Greeks as astronomers is based on the way in which they worked out the first great problem of astronomy, namely to devise a theory of the manner in which the heavenly bodies move which should account for the motions which we actually see. It had doubtless long been known that the stars moved as if they were points on a sphere whose axis passed through the pole-star* and which rotated once a day. All the heavenly bodies partook of this general motion, but some of them also moved relatively to the stars. Thus

* Not the present pole-star, for the direction of the earth's axis slowly shifts.

although the moon rose and set daily, it was seen each night to be in a different position relatively to the background of stars. It had long been known that there were seven bodies, called planets, that shifted their places among the stars. They were the sun and moon, Mercury, Venus, Mars, Jupiter and Saturn. The tracks of these latter five presented an intriguing problem. Though they share the general motion of all the stars, for most of the time they slowly fall behind them, yet sometimes keep pace with them for a while and for a time gain on them. Fig. 17, for example, shows the tracks of the planet Mars among the stars over a period of six or seven months.

The Greeks sought to find out how the planet really moved and to propound a theory that would enable an astronomer to predict its motion along this looped track. They were very well equipped for this task by their grand

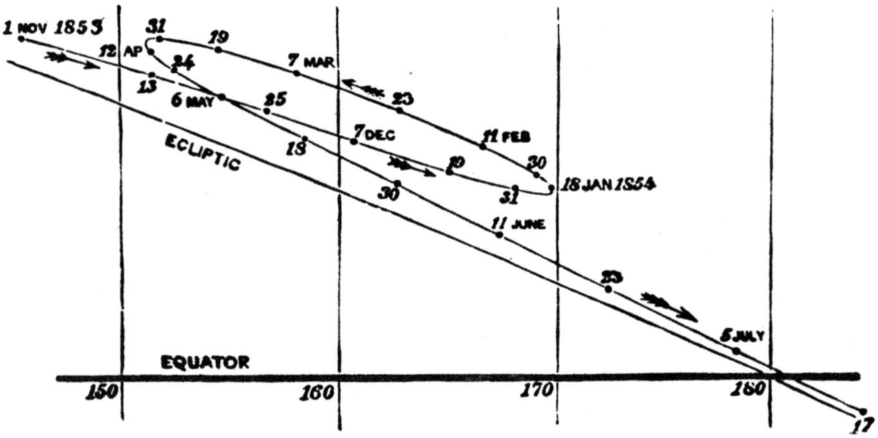

Fig. 17. The looped track of the planet Mars.

passion for geometry, and by the time of Aristotle they had carried the geometry of the sphere about as far as it could go. As astronomers, however, they were handicapped by their profound conviction that heavenly bodies must move in circles. So all the Greek astronomers based their theories of how the planets moved on the idea that all these bodies possessed two or more simultaneous circular motions, and they figured out in different ways the various orbits and speeds of rotation to be assigned to the circles or spheres carrying each planet in order to predict where it would be seen at any given date.

It would be a very long business to describe all the ideas that the Greeks put forward to explain these motions, so here it will be enough to look at those that lasted. With one notable exception the Greeks followed the evidence of

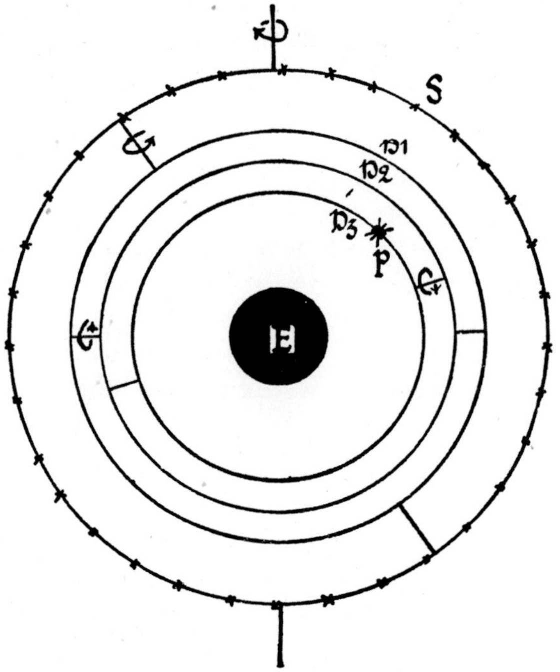

Fig. 18. The system of Eudoxus as applied to one planet. The circles represent sections of the various spheres, each of which rotates on axes fitting into the sphere outside it. The innermost bears the planet P.

their senses and assumed that the earth was the central body of the universe and remained motionless.

Aristotle followed two earlier astronomers (Eudoxus and Callippus) who tried to explain the motions of the planets entirely in terms of spheres rotating about the centre of the earth. He required fifty-five concentric spheres to move and carry the seven planets. Fig. 18 is a drawing of the mechanism needed to move just one planet. His system was to suppose three or four concentric spheres for each planet. The axes of each sphere fitted into the sphere outside it and each sphere turned at a different speed. The last sphere carried round with it the planet. This system made the planets, as seen from the central earth, describe a loop like an elongated figure 8, so as to accelerate, stop, retreat and go forward again, as they should. But this was not really the planet's path and so could never be made to predict the planet's position. Nevertheless the system was liked because it was so neat and symmetrical, with every motion circular and all the spheres concentric; so much so, that it was Aristotle's picture that persisted through the Middle Ages, though it was Ptolemy's system that was used by professional astronomers.

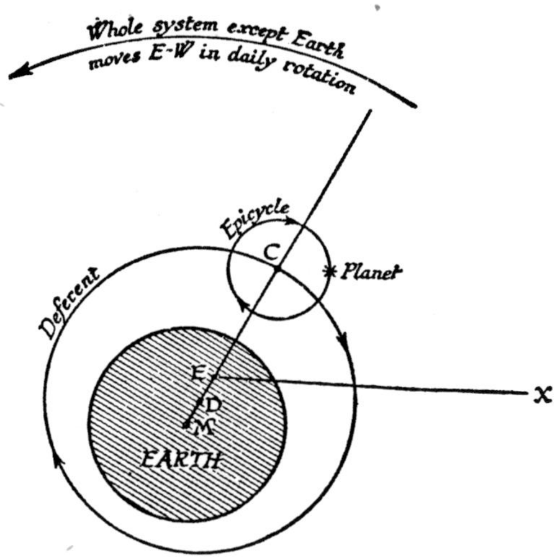

Fig. 19. The system of Ptolemy, as it applied to one planet. The planet revolves about the point C, which itself revolves about the point D, near but not at the centre of the earth. C does not move uniformly, but so as always to lie on the line CE which rotates uniformly about E.

This system is known to us from the works of Ptolemy, who flourished about A.D. 150, but it was probably invented by Hipparchus about 150 B.C. In this, as in Aristotle's system, the earth is at the centre of the universe and motionless. Again we will consider the path of only one planet at a time (Fig. 19). Each planet travels round a small circle (epicycle), whose centre travels round a large circle, whose centre is *near* but not *at* the centre of the earth. The centre of the small circle does not move at a uniform speed but so as to keep on the radius of another circle (deferent), which radius rotates uniformly about another point (equant). To account for the motions of some of the planets the centres of these circles had to be supposed to rotate in other circles! It is easy to see that if the speeds of the two main circular motions are rightly chosen, then the planet, as seen from the earth, will sometimes travel clockwise about it and sometimes anti-clockwise; it will in fact appear to describe loops, approximating closely to what is seen in the sky. This system worked pretty well; that is to say, it would predict heavenly phenomena with an error of perhaps an hour in some cases and a month in others! So we can call Ptolemy's theory a limited success, because it performed what it was asked to perform—the very impressive feat of predicting where the heavenly bodies would appear at any future time. It was not until the seventeenth century that its suppositions about the motions of the planets were found to be seriously inconsistent with the astronomers' observations so that it had to be abandoned in favour of the system of Copernicus, as modified by Kepler.

Fig. 20. A scaph consisted of a hemispherical bowl from the base of which rose a vertical needle (gnomon) reaching to the centre. The interior was graduated with lines, enabling the angle of the shadow of the gnomon to be read off.

The Polish priest, Copernicus, who, in the years before 1543, adopted the theory that the sun was at the centre of the universe and that the earth and other planets (except the moon, which he declared to be a satellite of the earth) rotated in circles about it, had been anticipated by a Greek astronomer, Aristarchus of Samos, who propounded exactly the same theory, about 270 B.C., but the men of his time could not bring themselves to suppose that the earth moved and the sun stood still, and his system was soon forgotten and had to be worked out afresh by Copernicus.

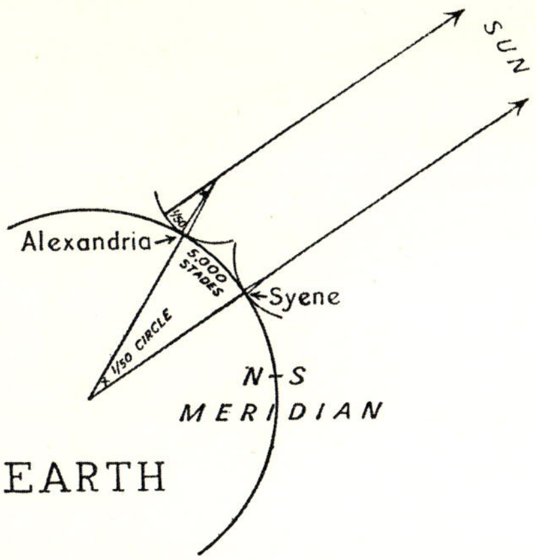

Fig. 21. How Eratosthenes measured the circumference of the earth.

In this system the apparent looped track of the planets depends on the velocities and directions in which they are moving about the sun. A tree viewed from a motor-car travelling round a circular track would seem to move backwards and forwards relatively to the horizon. The earth's orbit is the track: the tree the planet: the horizon the stars. The fact that the planet is also moving does not affect the issue, so long as it is not moving at the same angular velocity as that of the earth.

The Greeks' astronomy was by no means limited to solving the problem of the motions of the planets. They calculated the circumference of the earth from measurements and obtained results varying from 24,000 to 21,000 miles. The method was a very pretty one. Eratosthenes started from the supposed fact that at the moment when the sun was directly overhead at Syene (Assouan in Egypt), so that a vertical rod cast no shadow, at Alexandria (supposed to be due north of it) such a rod cast a shadow that made an angle of 1/50 of a circle with the vertical. This he observed with a kind of sundial, called a *scaph*, a hemispherical bowl with a vertical pointer at the centre and graduations for degrees.

Now he knew that Alexandria was 5,000 stades north of Syene, so it followed that 1/50 of the circumference of the earth was 5,000 stades. Thus a very simple calculation showed the earth to be 250,000 stades in circumference, which is just about 24,000 miles. A wonderful result—but Alexandria is not due north of Syene, nor 5,000 stades from it, nor would the angle made by the shadow have been precisely 1/50 of a circle. Eratosthenes was lucky in that his several errors cancelled out; but he was also a great man, not only to think of measuring the earth but also to do it. The Greeks even measured the relative sizes and distances of sun and moon and earth: their method was correct, but as their observations were inaccurate the results were a long way out. Yet in this way Aristarchus proved that the sun was far bigger than the moon or earth and an exceedingly long way off, which was a very important thing to know. Another great discovery of the Greeks was that of the precession of the equinoxes, though they did not discover its cause. The astronomical work of the Greeks may be summed up by saying that they founded the theory of astronomy so well that no one could improve on it for fifteen hundred years.

The Greeks did more for astronomy than for physics, chiefly because the former seemed to be a nobler science and the latter depended upon the making of what appeared to them to be trivial experiments. Nevertheless they made a beginning, especially where physics touched on mathematics. The founder of physics seems to have been one of the earliest Greek philosophers, Pythagoras, a very odd and legendary figure. His central idea, which may perhaps have been derived from the Mesopotamian culture, was that all things were composed of numbers; and so the Pythagoreans sought to find simple ratios of numbers in natural phenomena. He (or perhaps one of his followers) hit on the remarkable fact that when a stretched string was fingered so as to give notes of a recognized musical interval, the lengths of the sounding portion of the string bore the ratio of simple numbers. To put it at the simplest: halve the length of the string and it will sound the octave; make its length three-quarters and it will sound the fourth. This is the first recorded physical experiment. It impressed the Greeks vastly and the echo of it is heard in Sir Thomas Browne's description of music as 'the mystical mathematics of the city of heaven'.

The practice of experiment increased greatly after Greece was conquered by the Macedonians and the centre of learning shifted from Athens to other Greek-speaking centres and especially to Alexandria. The first great scientist of this period was Archimedes (287–212 B.C.). That he was a great mathematician is not so remarkable as the fact that he performed physical experiments. He invented the science of hydrostatics, explained the lever, and his 'principle' is still taught to the budding scientist as the fundamental means of determining densities; he was also the inventor of the Archimedean screw which has been used for pumping water even up to modern times (Fig. 23).

Archimedes was, however, rather ashamed of working with his hands. We are told that, although at the siege of his city, Syracuse, by the Romans, he invented such war-machines that the besiegers dared not touch a loose rope-end in case it should be a trap, yet he thought it unfitting for a philosopher to record such things. His method in physics was, it seems, first to experiment and so find out the facts about, let us say, the position in which bodies floated or hung when suspended, and then, concealing all this, to invent a beautiful mathematical deduction of the results from first principles, which was all that he published. His experiments were but a scaffold erected to help him build his mathematics.

The atmosphere of Alexandria was very different from that of Athens. The city was a vast commercial centre, in which met and mingled all the races and cultures of the near East. Science flourished in its great Museum and Library, but those who practised it were on the whole more interested in facts and less in philosophy than the classical Greeks. Accordingly we find that

Fig. 22. Pythagoras, on his visit to Egypt, demonstrates the relationship between the pitch and length of strings. For the circumstances of this picture, see Appendix I.

Fig. 23. An Archimedean screw, as used for raising water in modern Egypt.

Fig. 24. Philo of Byzantium investigates the properties of air.

as time went on the Greek-speaking people showed more and more interest in physics and especially in what we call gases and they called 'spirit' or 'breath': probably because their medical men held that the motions of the body were actuated by means of spirits or breath. There survive two interesting treatises which cannot be dated very certainly, that of Philo of Byzantium and Hero of Alexandria. Philo seems to have flourished about 200 B.C. Only a fragment of his work on gases has survived and that in an Arabic translation. But this is enough to show us that he had thought about gases and done experiments upon them and their expansion by heat. He describes the first experiment on this subject. A leaden ball was connected by a pipe to a vessel of water: when it was heated by the sun air bubbled out through the water, and when the ball grew cold again the water travelled up the tube into the ball. Philo explains it as due to the 'thinning-out' of the air by fire. He is the first, too, to do the grand old experiment of burning a candle in air confined over water, the demonstration of which still begins our school chemistry courses. His explanation of the rise is that some of the air 'perishes' or 'is consumed' by the action of the fire. Eighteen centuries were to pass before anyone thought of a satisfying explanation.

Another writer on gases, Hero of Alexandria, is thought by some to have lived 150 years before Christ, by others to have lived 250 years after. We should not have much difficulty in saying whether an English book was written in 1500 or 1900, and the fact that the scholars cannot guess the date of Hero's within 400 years shows how much more static was the ancient world than ours.

Hero was evidently a considerable mathematician, engineer and physicist. Most of his works are lost, but his treatise on pneumatics is preserved and seems to us to display a very queer state of mind. The book is all about what we would call conjuring tricks or illusions. Thus it describes and pictures a sort of penny-in-the-slot machine for providing holy water; a very large number of trick-vessels designed to pour or fill or empty in unexpected ways, or to give wine and water at will. There are all kinds of automata, figures that move automatically, birds that sing by compressed air; quite a considerable section is devoted to the blowing of small organs by means of air displaced by running water. Typical of Hero's ingenuity is a complicated contrivance intended to make the doors of the temple open when a fire is lighted on the altar. The altar is a copper box, the fire makes air in it expand, the air drives out water from another vessel beneath the floor into a bucket, the bucket descends and pulls on ropes wound round the axles of the doors! The worshippers were doubtless much impressed. We may feel rather superior to all this trickery, but not only do these contrivances show the author's understanding of hydraulics and mathematics, but they also bear witness to

Fig. 25. The contrivance, figured by Hero of Alexandria, by which the doors of a shrine open mysteriously when a fire is lighted on the altar.

Fig. 26. Hero's reaction-turbine, the first steam-engine, was intended as an amusing toy, not as a source of power.

a state of craftsmanship in which gas-tight vessels, pipes, taps, pistons and so forth, could be constructed.

It is in this company that we find the first steam-engine. Water is boiled in the hemispherical cauldron, the steam escapes through the jets and turns them by reaction. There is no provision for driving anything by its aid. It

Fig. 27. St. Augustine of Hippo surprised and disturbed by the powers of the loadstone.

was, in fact, intended only for an amusing toy, and it was not till the sixteenth century that anyone wanted to, let alone was able to, make it useful.

The Greeks and Romans knew of electricity and magnetism, in so far that they knew that amber (the Greek word for which is *electron*), when rubbed, attracted light bodies, and that a black stone, which they called *magnes*, had the strange power of drawing iron. The loadstone, magnes, intrigued them greatly. Thales, the first Greek scientist (600 B.C.), thought it had a soul because it drew the iron to it. Lucretius, the Roman poet, tells us he had seen iron filings 'boil' in a copper basin when the loadstone was passed beneath it. But the most interesting account is that of St. Augustine—not the saint that converted England, but St. Augustine of Hippo, greatest of Christian philosophers and theologians. He tells us in his book *The City of God*, written between A.D. 413 and A.D. 426, how astonished he was at the loadstone. 'When I saw how the stone snatched at the iron, I shuddered all over.' He goes on to give the first description of induced magnetism. 'For I saw an iron ring seized and suspended by the stone; then, as if the stone had given a share of its own power to the iron it had seized, that same ring moved to another and suspended it: as the first ring clung to the stone, so the second ring clung to the first: in the same way a third and a fourth, so as to make a chain, not a ring within a ring, but clinging to each other's surfaces.' And he says that a friend of his had been shown by a Moorish prince how a magnet held under a silver plate would move iron on the other side of it. The fact that the magnet could move something at a distance or through a metal plate astonished everyone. The magnet and amber were in fact the only examples of attraction known to the ancients. The notion of action at a distance was abhorrent to them: it did not fit into their world.

Interested in the loadstone as they were, the ancients did not discover its north-pointing property, much later utilized in the mariner's compass, which was one of the several mysterious discoveries of the Middle Ages which cannot be attributed to any known person. It is thought that the Chinese knew of it before the Western world, but even this is uncertain. Sailors seem to have known of the loadstone not long before A.D. 1200. The simplest way to use it is to put a piece on a float, whereupon it will swing round so that its poles lie in the north and south line. The next step was to magnetize a needle by 'touching' it with a loadstone and stick it through a piece of wood or a straw. It seems that the sailors of the early thirteenth century did not keep a compass but made one in this way when they wanted it—that is to say, in a fog or when the sky was obscured by cloud and they were out of sight of land: but even before 1300 they seem to have had pivoted compasses, not very unlike modern ones. It was one thing to know that the magnet drew the iron, but quite another to explain it. The thirteenth-century answer was that the magnet had an attractive virtue! If that is no answer, it is all the answer men could

Fig. 28. The primitive compass, a magnetized needle stuck in a straw.

give before the nineteenth century, and I doubt if much more can be said today.

The later Greeks made another notable advance in scientific practice, the invention of the chemical laboratory and the art of distillation, that is to say, the evaporation of a liquid and the condensation of its vapour. We cannot name the inventor of distillation. True, Aristotle says, 'salt water when it turns into vapour becomes sweet and the vapour does not form salt water when it condenses again': but we do not think he had a real still. The first alchemists—men and women who sought to make precious metals from base metals—possessed very good stills about A.D. 100 and we do not hear of stills for any other purpose for several centuries thereafter. We cannot trace the first inventor of this instrument: Maria the Jewess, an alchemist of this period, is

Fig. 29. The alchemist, Maria the Jewess, in her laboratory at Alexandria, about A.D. 100. The three-armed still, in the centre, was her invention; the other apparatus is such as was used by the alchemists of the period.

recorded as the inventor of the three-armed still shown in Fig. 29, but we cannot regard her as the first to make a still.

Fig. 30. The Alembic. Such a piece of apparatus might have been used in any century from the first to the nineteenth.

The typical still which remained in use till the middle of the reign of Queen Victoria, as first illustrated by these early alchemists, is shown in Fig. 30. It is an excellent instrument for slow distillation, and deserved its long life of perhaps 1,750 years. The material to be distilled is placed in the 'body', into which the 'head' which serves to condense the vapour is fastened by some kind of adhesive, such as flour and water: the vapour rises, condenses on the dome of the head, runs down the glass into the gutter and flows out through the 'beak' or 'nose' into the receiver. What was the significance of distillation to the alchemists? They thought of matter as being like man, and therefore as having a body and a spirit. By distillation they thought to isolate the spirits of things, which

Fig. 31. A still with a fractionating column illustrated in a pharmaceutical work of 1512 by Hieronymus Brunschwyg.

Fig. 32. The first professional chemists at their work of analysing precious metal and gold ores.

should possess their characteristic activity and be the seat of their special properties.

Their laborious distillations did not bring about their end, the making of gold, but, John Donne tells us,

And as no chemic yet th' elixir got,
But glorifies his pregnant pot,
If by the way to him befall,
Some odoriferous thing or medicinal.

So in fact the alchemists and those they taught discovered by distillation some notable human amenities. First they learned to distil perfumes; then they discovered the distillation of alcohol from wine. This was not apparently recognized at first as something to be drunk, but rather as an incendiary material, a water that burned; from the thirteenth century, spirits were recognized as a medicine for old age, and only in the sixteenth century as a convivial

Fig. 33. Incendiary bombs fired from cross-bows at a mediaeval town, while fire-watchers attempt to throw them down.

drink. The same process, applied to mixtures of alum, vitriol and saltpetre, yielded aqua fortis (nitric acid), which was soon produced on a large scale as a means of separating silver from gold.

The Greek alchemists did not confine themselves to distilling, for their laboratories contained all manner of flasks, crucibles, funnels, stirrers, sand-baths, water baths, furnaces, etc. They were indeed the first laboratory workers, but they were scarcely to be called chemists in the modern sense, for their interest was not in discovering the constitution and properties of matter, but only in their supposed art of making gold.

The later Greek chemists also invented incendiary bombs which were composed of pitch, resin, petroleum, sulphur, etc., for siege and naval warfare. This 'Greek fire' was a formidable weapon, and was much in use throughout the Middle Ages.

We have pictures of troops (*c.* 1450) discharging incendiary bombs from

Fig. 34. The 'boy's trick' of letting off a cracker, practised upon Roger Bacon.

crossbows on to roofs and of fire-watchers throwing them off again. Incendiary mixtures were soon supplemented by gunpowder, a mixture of charcoal, sulphur and saltpetre. This is another mysterious invention. The Chinese had used firework mixtures containing saltpetre, and sometime near 1250 saltpetre found its way to Europe. Roger Bacon, the English scientific friar, writes of powder, about 1269, as a filling for crackers, but it was certainly after his time that it was first used in guns. He writes in his *Greater Work*:

> 'And we have experience of this from that boys' trick which is done in many parts of the world, namely that by a contrivance no bigger than a man's thumb and the violence of that salt which is called saltpetre, so horrible a noise is made by the bursting of such a little thing, a mere bit of parchment, that it seems to exceed the loudest thunder and in its brightness surpasses the biggest flash of lightning.'

We may guess that Roger Bacon is relating his own experience. What a blessed age was that in which a cracker gave the loudest bang that anyone had heard!

Throughout human history the physician and surgeon have striven to cure human ills, and have slowly gained a knowledge of remedies for disease and injury and of the structure and working of the body. At every period the physician had a huge armament of drugs, and numerous manuscripts describe his materia medica and the means of compounding them. Many of these drugs were effective purges, vomits, irritants, narcotics and the like, but it cannot be said that they cured the diseases to which men were subject. The body sometimes overcame the disease and sometimes did not, while the physician took the credit or the blame. Nevertheless it must not be thought that medicine made no progress before the age of modern science. A great many diseases had been recognized and the medical man could at least reassure or warn the patient's relatives. Something was understood concerning the aiding of nature in her task: the use of rest and diet were at least appreciated. If the routes of infection were not known, the danger was familiar. In the domain of surgery, even in very early times, there were some effective procedures—setting fractures, reducing dislocations, opening abscesses, operating for stone and the like, but the absence of anaesthetics and the invariable sepsis of surgical wounds, so far restricted the surgeon's field, that his ministrations made little difference to the prevalence of disease and the high mortality. But even that small difference was precious. Cures, were they nature's or the surgeon's, gave hope to the sick. A man who had an injured part burned with a red-hot iron was physically none the better, but he had made the great effort of consenting to it, and gained a corresponding confidence in his ultimate recovery.

Fig. 35. A pharmacy of the fifteenth century.

Fig. 36. A Greek clinic. The surgeon is about to bleed the central figure. The injured and deformed are waiting, as today, for their turn.

The Greeks and their followers were extremely interested in the structure and working of the body, but the problem was much too difficult for them. There was a prejudice against dissection of human bodies, but even the dissection of animals was not very carefully performed, if we may judge by the results. The ancients came to know in a general way what the various organs did, but understood almost nothing about how they did it. The science of the working of the body, which we call physiology, is largely dependent upon a knowledge of chemistry, which itself scarcely began to be a science before the seventeenth century.

So much then for the first beginnings of science. What men had learnt by the year 1600 of the nature and behaviour of things was a great advance on the ignorance of primitive man. It was enough to give mankind an intelligible and uplifting picture of the world: nevertheless, much of the picture was untrue and scarcely any of it was exact; and indeed the extent and depth of the knowledge of nature that man had gained in 5,000 years was almost negligible compared with what he was to gain in the next 350.

CHAPTER TWO

EXPERIMENTAL SCIENCE BEGINS

BY THE END OF THE Middle Ages men's knowledge of the world and industrial technique had outgrown what the ancient science had to tell. What was written in the books was not enough for them and they began to trust their own eyes. Thus at this time the German astronomers began to make new star-maps, to record the positions of the planets with greater care; moreover they were able to publish their findings, for printing came into use in the latter years of the fifteenth century and thus enabled the ideas and observations of scientists to reach the rapidly widening circle of those who could read and write. No more than careful observation was needed to show that a better theory of the planets than Aristotle's or Ptolemy's was wanted, and in 1543 the Polish priest, Copernicus, published his theory that the sun was stationary, while the planets and the earth (itself a planet) rotated about it, and the moon rotated about the earth. The stars, Copernicus supposed, were motionless and their apparent daily circling about the pole was simply due to the earth's rotation. This was no more than Aristarchus had said, but his works had long been lost. Copernicus knew only that some of the ancients had thought on these lines: he worked out the theory again, and in a much more thorough way. Hardly anyone believed him at first, and it was not until some sixty years later that the great controversy between those who held the earth to be stationary at the centre of things and those who gave that position to the sun began to shake the learned world.

It was not only the astronomers who began to trust their eyes. In another field, that of anatomy, surgeons and artists began to dissect, observe and draw. The Italian artist, Antonio Pollaiuolo, about 1540, may have been the pioneer, but Leonardo da Vinci's are the first of the new anatomical drawings that we possess. His drawings were not published, but in the same year (1543) as Copernicus published his theory, there appeared a wonderful volume *On the Fabric of the Human Body* by the Fleming, Andreas Vesalius. Here was an attempt to draw the vessels, bones and muscles as they really were: the work has its errors, of course, but is incomparably better than anything that had previously been published.

At the same period there arose a wide general interest in industries and trades, formerly the preserve of the illiterate; beautifully illustrated books on mining, fortification, metallurgy and machines began to appear, and scientific men even tried to design industrial plant, though not with much success.

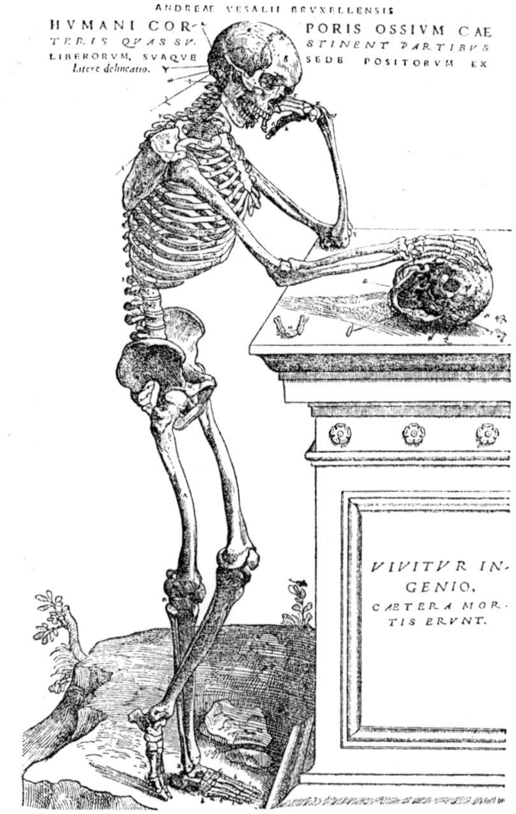

Fig. 37. The skeleton contemplates mortality (Vesalius)

Thus the men of the sixteenth century became interested in the things that science studies, but they did not yet know how to go about the task of finding out the workings of things. The man who taught the world to do this was Galileo Galilei, who may be called the founder of experimental science.

We may sum up Galileo's long life of scientific work by saying that he made it his task to confute the science of the Greek philosophers and to found a new science on observed fact and nothing else. He was born in the year 1564 (the same year as Shakespeare) and died in 1642, the year in which Isaac Newton was born. Even as a boy, Galileo was passionately interested in mathematics and physics and he spent a long life on nothing else but science. He was perhaps the first full-time scientist of the modern world. A clever craftsman as well as a great scientist, he made his own apparatus, ground his lenses, and was for ever thinking out and making ingenious contrivances to prove his points and answer his questions. Unlike the modern scientist he did not publish many of his discoveries until long after he had made them; accordingly we cannot review his work in chronological order, and must group his manifold discoveries according to their subjects.

At the bottom of Aristotle's whole theory of the universe were his ideas of motion; that bodies moved 'naturally' to their place at a speed that depended on their weight and the resistance of the medium; and that bodies which moved in any other way had to be kept in motion by something. Galileo seems to have doubted these 'laws' from the beginning. As we saw, Simon Stevin had disproved Aristotle's ideas about falling bodies, but it is unlikely that Galileo knew that he had done so. He seems to have done much the same kind of experiment though he does not tell us when or where. He certainly more than once dropped two different weights from high places and noticed that the two weights did not reach the ground exactly together, if the height was great; he thus observed that the velocities of a 10-pound and a 1-pound weight freely falling through air were certainly not in the proportion of ten to one (as Aristoteleans supposed), but perhaps more like ten to nine. But Galileo went much further than Stevin, who had only disproved Aristotle's views, in that he set to work to find out how falling bodies really did move. Aristotle had said nothing about acceleration; but one or two learned mediaeval philosophers had argued that the 'heaviness' of a falling body continues to act throughout its fall, and that consequently a falling body ought to move faster and faster. They even worked out the right law of motion, but they were unable or unwilling to test it and find out for themselves whether they were right or not!

Galileo tried to observe the acceleration of falling bodies. But how was he to do it? There were no stop-watches or chronographs, but only very crude and inaccurate blacksmith's clocks without minute hands, as a rule—let alone second hands. So instead of trying to time anything moving as fast as a metal ball dropping from a tower, he timed metal balls rolling down slopes. He saw that, if the slope was increased until it was vertical, then rolling became identical with falling, and so the same law ought to apply to rolling and falling. So to measure the speed of balls rolling down a slope he designed a piece of apparatus—a novel thing to do in those times.

He made a beam twenty-two feet long, set edgeways so as not to bend, and having a groove in which a brass ball could roll. How was he to time it? For this he went back to the old Egyptian water-clock—a bucket with a hole in it; he weighed the water that flowed while the ball was running. He measured the time the ball took to run, first, the whole beam, and later, various fractions of it. Thus he found the ball took *half* the time to run a *quarter* of the distance: and by a series of experiments he showed that the distance run by the ball was as the square of the time needed to run it. In other words *distance ÷ square of time* was constant for a given slope. This result could not be accounted for by any other hypothesis except that the ball was uniformly accelerated while it ran down the slope.

Galileo thus proved that Aristotle's ideas about the natural motion of heavy bodies were wrong: but what of his idea of their 'violent' motion? Aristotle supposed that when someone shot an arrow or threw a stone something kept it in motion: indeed he seems to have supposed that the air which was pushed away from the front came round and pushed the arrow from the back! The crucial experiment would have been to try to project a body in a vacuum, for according to Aristotle it was impossible to throw a stone horizontally in a vacuum, because this could do nothing to keep it in motion. But neither in Aristotle's time nor Galileo's could this experiment be done, for no one in the world had found a way to make a vacuum. Yet throughout the Middle Ages many philosophers had disagreed with Aristotle and had thought that the

Fig. 38. Galileo lets a ball roll down a groove in an inclined beam and times it by weighing the water that spurts from a bucket during the run.

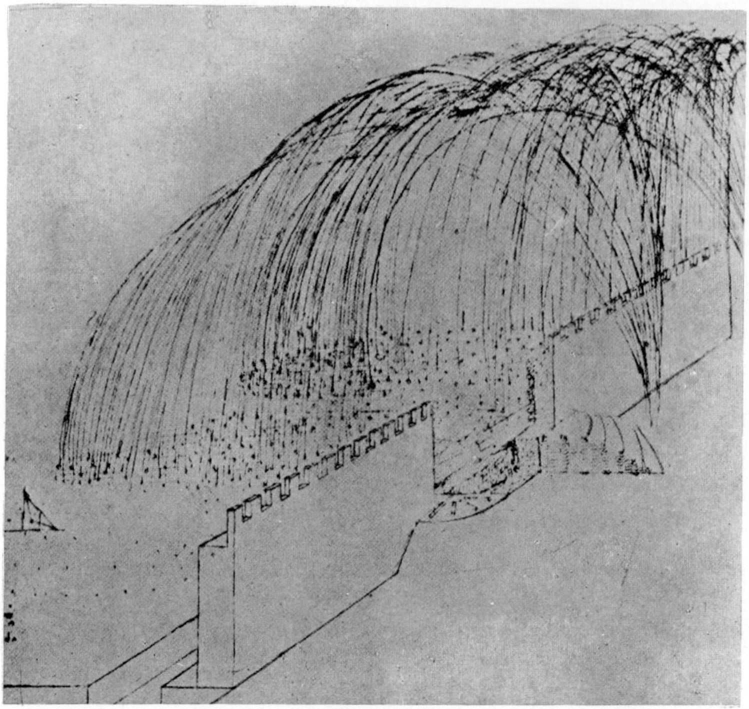

Fig. 39. Leonardo's drawing of the course of a shower of projectiles, a close approximation but without a theoretical basis.

moving arrow, or what you will, had something in itself, a sort of quality that kept it going, which they called 'impetus', much the same idea as our 'inertia': Galileo held similar views, though he said nothing of qualities and just expressed his opinion that if nothing stopped a moving body it would go on for ever. He asked us to imagine a smooth plane sloping never so little downwards. On that a moving ball would accelerate. If it sloped never so little upwards the ball would decelerate. So he said, if the plane were level and smooth and the sources of friction were imagined away the ball would go on for ever. This experiment could not, of course, be performed: we may call it a 'thought-experiment'.

Now these two ideas, first that a body proceeding downwards was accelerated, secondly that a body proceeding horizontally kept a constant speed, enabled Galileo to solve a problem that had long been a puzzle—namely to specify the curve followed by a projectile. Some thought a cannon-ball went straight out of the cannon, then after a time began to curve downwards, then dropped sheer: others such as Leonardo came very close to the true curve, but no one had specified it in mathematical language. But Galileo saw that in each second the projectile travelled the same distance horizontally (like the

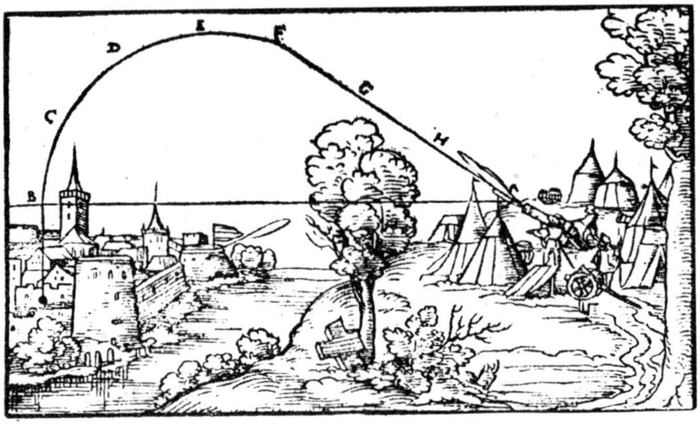

Fig. 40. The erroneous view of the trajectory of a cannon-ball as figured by Walter Ryff (1540)

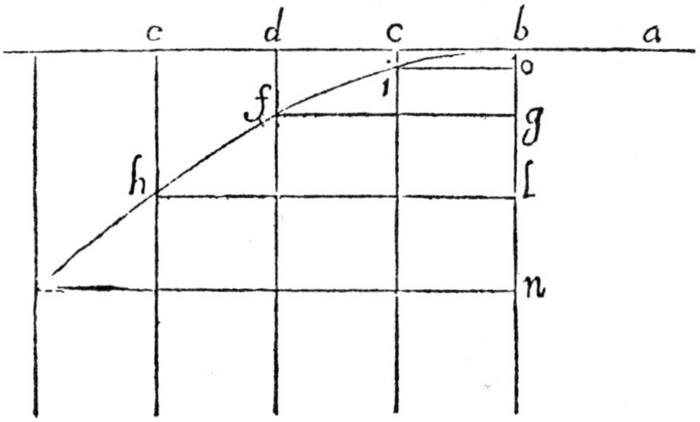

Fig. 41. The parabolic course of a projectile fired horizontally, as mapped out by Galileo on theoretical grounds.

imaginary ball on the slab) but at the same time it fell through distances, which were as the square of the time of fall, e.g. 1, 4, 9, 16, 25, etc. The curve which fulfilled these conditions had been known since the time of the Greeks and was a parabola. His figure is reproduced as Fig. 41.

Galileo saw that it was very important to investigate air and, if possible, to make a vacuum. Aristotle considered that air had no weight and Galileo set out to prove the contrary. Since air pumps had not been invented, he could not weigh a flask, first evacuated, then full of air. But he thought of a very pretty plan. He fitted a bottle with a leather valve and tube. He forced water into it so as to compress the air; he then weighed the bottle. Next he released the compressed air by pushing down the valve and weighed the bottle again. The volume of air that escaped from it was clearly the same as that of the water pushed in. So the loss of weight that took place when the pressure was released was the weight of that much air. So it was proved that air had weight: in fact Galileo made the weight of the air about twice what it should be: probably because he lost a droplet of water in releasing the valve.

Galileo did his best to obtain a vacuum and he quite possibly did so by filling a cylinder completely with water and pulling out the piston (Fig. 42). But this experiment was not enough to convince the world and the feat was accomplished after his death by one of his friends and pupils, namely Torricelli.

Fig. 42. Galileo attempts to prove the possibility of a vacuum and measure its 'force' by pulling the piston out of a cylinder filled with water.

Fig. 43. Torricelli demonstrates the vacuum. When the upper ends of the tubes are raised to the vertical, the mercury, which originally filled them, remains at approximately thirty inches above the surface of the mercury in the bowl, leaving an empty space, which when the tube is lowered once more becomes filled with mercury.

Fig. 44. Otto von Guericke's first attempts to evacuate vessels.

He filled a tube, about a yard long, with mercury and held it vertically with its opening under the surface of mercury in a bowl. The mercury in the tube fell until it was some thirty inches above the surface of the mercury in the bowl (Fig. 43). If the tube was then inclined so that its end was less than about thirty inches above the mercury in the bowl, the mercury rose to the top of the tube and occupied it completely. There had been nothing else but mercury in the tube, nothing could get through the glass, so the space above the mercury contained nothing—and was a true vacuum.

What kept the mercury up? The answer seemed to be the pressure of the air, and to prove this the apparatus was taken to the top of a high hill where there was less air above it. The mercury now fell to about twenty-seven inches, a fact which proved that it was the air that held it up. Such simple experiments really, yet they proved the reality of the vacuum and demonstrated the huge pressure of the atmosphere which nobody had yet realized.

It was rather difficult to study the properties of the vacuum at the top of the tube, so it was a great step when Otto von Guericke, Burgomaster of Magdeburg, a rich amateur scientist, made the first air-pump in 1650. His first plan was to pump the water out of a full barrel: but air sizzled in between the staves. So instead of the barrel he used a copper globe, but this collapsed with a loud report. At the third attempt, however, he constructed a pump which would do the work.

His most famous experiment to show the pressure of the air, was that of the Magdeburg hemispheres. He made two hemispheres of brass, about eighteen inches in diameter, with flanged edges that fitted exactly. He pumped out the air and found that a team of sixteen horses could not pull them apart (Fig. 45).

Robert Boyle and others improved upon von Guericke's pumps. Scientists soon became familiar with the ideas of gas-pressure and vacuum, so that by the close of the century they were already beginning to put them to use in the first steam-engines.

Another very important discovery made by Galileo when he was a young man was the fact that a pendulum of given length makes the same number of swings in a given time, no matter whether the breadth of the swing be large or small. Actually this is only approximately true, but it proved to be near enough to the truth to enable men to use pendulums to regulate accurate clocks. Galileo is said to have made his discovery by watching a lamp swinging from the roof of the Cathedral of Pisa and timing the oscillations by the beats of his pulse. It was not a very accurate way of timing and we do not doubt that he later checked his results against better timekeepers than the human heart. Galileo designed a pendulum-clock, but it does not seem to have been constructed, and the idea lay fallow until Christian Huygens applied it in a rather different manner.

Fig. 45. von Guericke's demonstration of the force needed to separate two hemispheres enclosing an evacuated space.

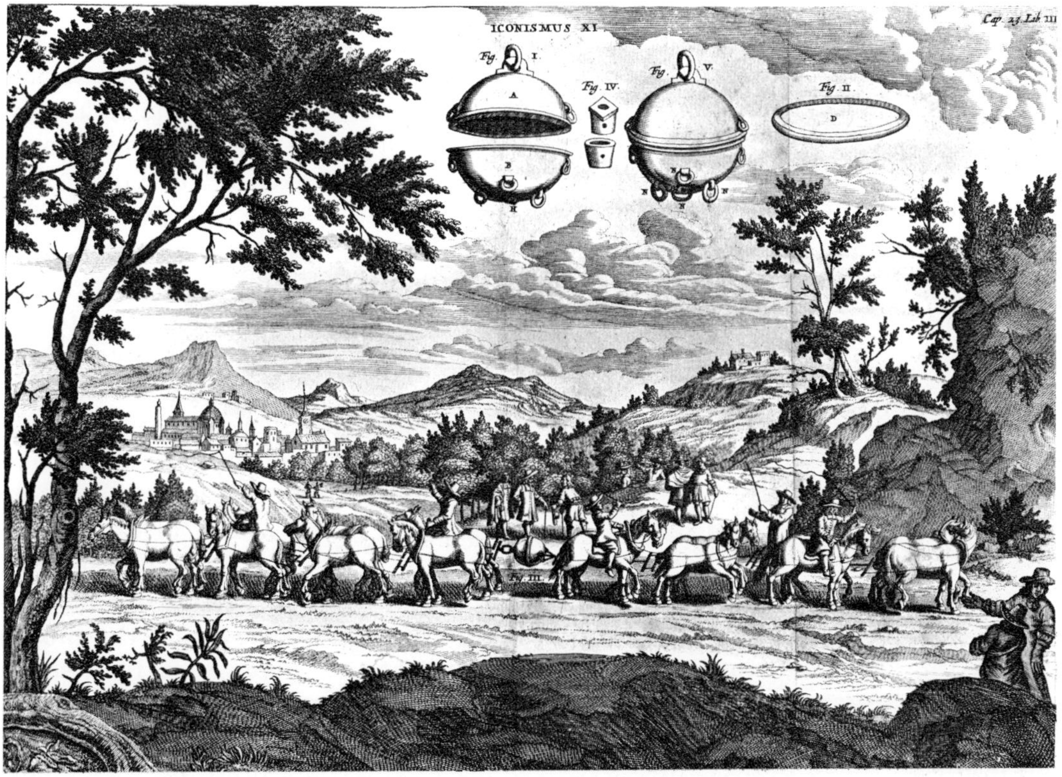

Galileo seems to have been the first to use a thermometer for showing changes of temperature. The instrument was simply a glass tube with a bulb at one end filled with air, the other open end being dipped into water. Galileo does not seem to have been much interested in it nor to have recorded the discovery until 1611, when a physician called Santorio published an account of a very similar instrument, which he used to ascertain the 'heat of the body'. Santorio had no fixed scale of degrees but judged the 'heat' by the rate at which the liquid fell. The air-thermometer was improved by fitting a scale of degrees, but the instrument could never be accurate because not only the temperature of the air but its ever-changing pressure affects it. It was therefore an important step when, in 1641, Ferdinand, Grand Duke of Tuscany, thought of measuring temperature by the expansion of a liquid in a sealed thermometer, in principle the same instrument that we use today. It was a wonderful step to be able to measure heat, but it took many years to devise a standard thermometer. It was not until 1693 that a reproducible scale was based on dividing into a number of equal parts the total expansion of the liquid between the temperatures of melting ice and boiling water: indeed this plan came into general use only in the years around 1720.

The probable reason why Galileo did not explore the consequences of this invention was that, after 1609, astronomy became his absorbing interest. He had for a long time been a believer in the ideas of Copernicus and no doubt he was anxious to confute Aristotle's ideas about the motions of the heavenly bodies. By a piece of good fortune he heard that a Dutchman had used lenses to make an instrument which caused far-off things to seem near. Given this hint, it did not take him long to devise a telescope, and as he was a very skilful worker with his hands he soon made an instrument that not merely magnified but gave a sharp image. Most men would have thought of the instrument as something useful to generals or ship's captains, but Galileo at once directed it at the skies. As he was already a skilled astronomer, he immediately understood the significance of what he saw.

Galileo's first great discovery was concerning the moon. According to the Aristoteleans, it should have been a perfectly smooth and geometrically exact sphere composed of the fifth element, unknown on earth. Galileo studied the boundary between the light and dark parts of the moon rather before it was half-full. Just inside the dark part he saw spots and lines of light, and he interpreted these spots and lines as the summits of mountains and the rims of valleys lighted up by the rising sun. From the analogy of dawn on the moon and on the earth, he leapt to the conclusion that the moon must be a solid rugged body—another earth and presumably made of the same materials. This notion, that the moon was another world, aroused enormous excitement and interest.

Fig. 46. Galileo as a very young man watches a lamp swinging from the roof of the cathedral at Pisa, and observes that the time of swing is constant and independent of the arc of swing.

Fig. 47. Galileo, who was living at Padua, visited Venice in 1609 to show his telescope to the Doge and Senate. He is here pictured as taking the opportunity of making a few celestial observations.

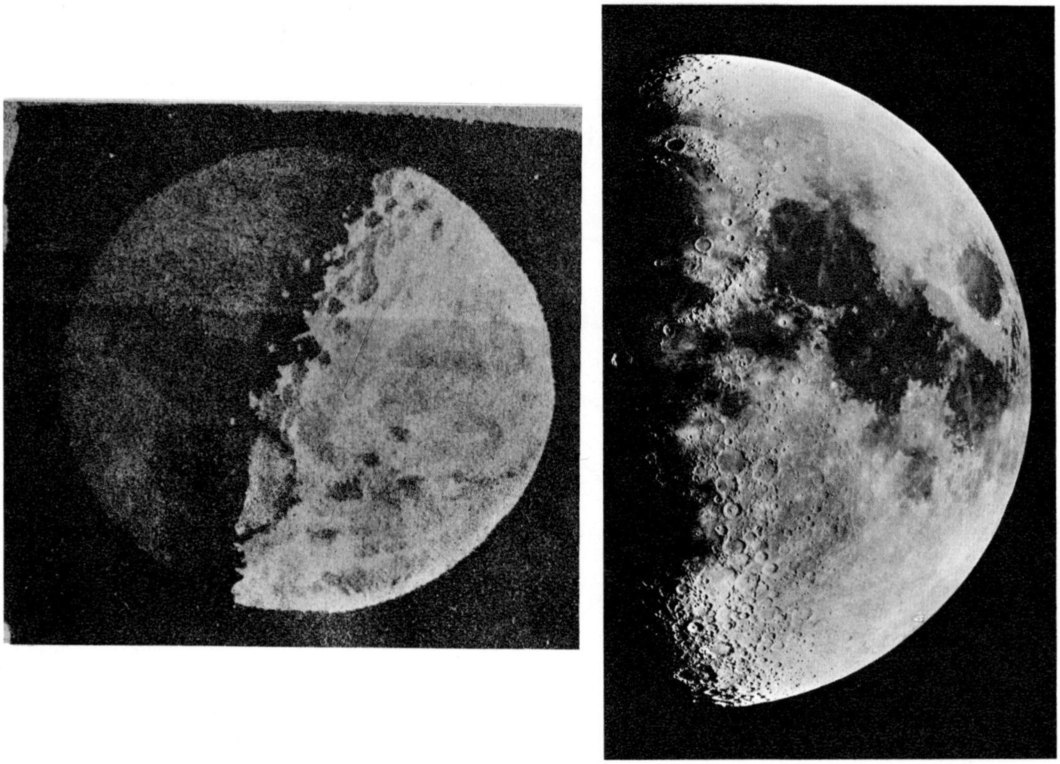

Fig. 48. *Left:* Galileo's drawing of the moon. *Right:* A modern photograph. Galileo's telescope showed him very much less than the modern telescope, but enough to refound astronomy.

His telescope was not powerful enough to reveal much about the planets but, when he turned it to the planet Jupiter four little stars were always seen to accompany it and to alter their relative positions on successive nights. Galileo soon proved that these little stars revolved about Jupiter, just in the same way as Copernicus had said that the moon revolved around the earth. Moreover, here were little bodies rotating round a big body; Copernicus had supposed that the little planets rotated round the big sun, not the big sun round the little earth; so these observations supported, if they did not prove, the Copernican system, which had formerly rested only on the fact that it gave a simpler account of the known universe.

It is probable (but not certain) that Galileo was the first to observe sun-spots. These showed not only 'imperfections' in that luminary but also changes —clean contrary to Aristotle's idea of a heavenly region where nothing came to be or passed away.

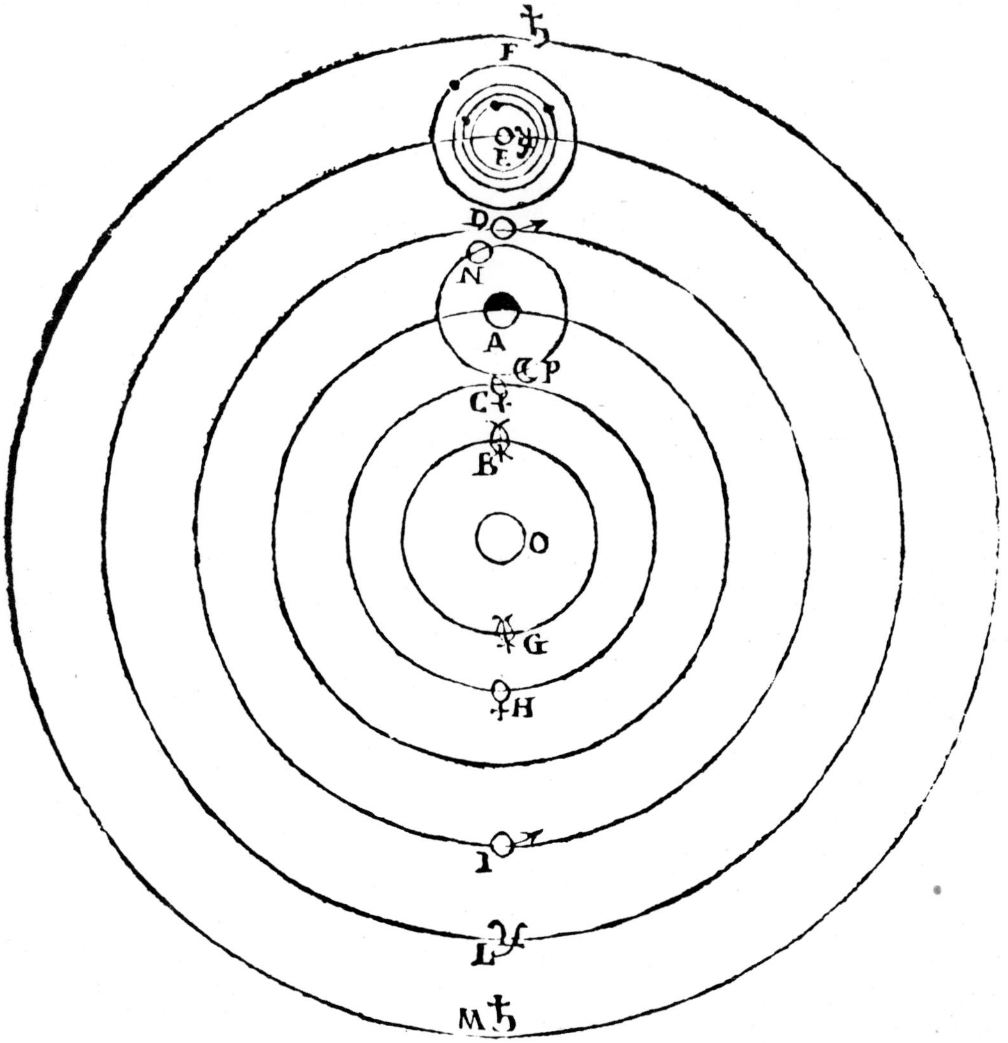

Fig. 49. Galileo's diagram of the solar system. Compare the mediaeval view (Fig. 12).

Next Galileo showed that the planet Venus showed phases like the moon and so he concluded that it must be an opaque body like the moon or the earth, not some sort of spiritual intrinsically shining being. These observations concurred to support the argument, 'If these planets, Venus and the moon, resemble the earth, why not agree with Copernicus that the earth, like them, is a planet and is not the unique centre of all things.' In Fig. 49 we see Galileo's picture of the Copernican system with the addition of his own discovery, the four satellites of Jupiter.

Galileo's advocacy of the Copernican system precipitated the question as to

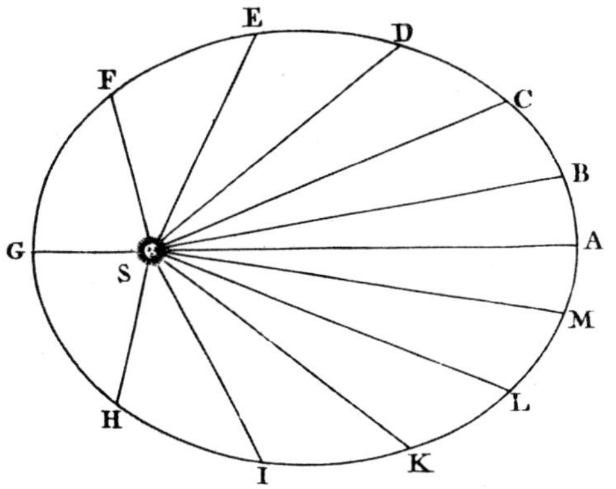

Fig. 50. To illustrate Kepler's laws. Each of the compartments of the ellipse is of the same area and is traversed by the planet in the same time.

whether the Catholic faith implied that the ancient astronomy was true. The problem had not been properly thought out and the final result of muddle and intrigue was that the Copernican system was held to be heretical. Galileo was forbidden to hold or defend it: it is a matter of controversy whether he was forbidden to teach it. At all events he did set it out in a very favourable aspect in his Dialogues of 1632. For this he was tried by the Inquisition and punished by confinement to his house for the last seven years of his life. The Catholic Church retreated, though rather slowly, from the position that had been taken by the Inquisitors; in Catholic countries astronomers had to speak with caution and the further progress of the theory of the solar system was made in the northern countries where the Church's authority held no sway.

Galileo was as convinced as Aristotle that the heavenly bodies had none but circular motions, but no one was able to make exact predictions of planetary positions on this basis. The most convincing argument for the idea that the sun was the central body of the solar system was Johann Kepler's exact prediction of the path of a planet. He had discovered that if the Copernican system were modified so that the planets described elliptical orbits instead of circular, their motions could be exactly predicted by mathematical laws. After a vast amount of observation and calculation, Kepler showed that the planets moved in ellipses, the sun being in one focus: and that the planet moved at a speed which varied in such a way that the line joining it to the sun swept out equal areas of the ellipse in equal times. Lastly, to his great delight he found that the squares of the planets' times of rotation were proportional to the cubes of their distances from the sun.

These laws were so exact that those of his contemporaries and successors that could understand him firmly believed that the earth and planets moved about the sun, though ordinary people did not make up their minds on the subject until after the time of Isaac Newton.

Copernicus had shown that the plan of a stationary sun and stars, and a rotating earth, revolving with the other planets about the sun, was simpler than that of a stationary earth and therefore more easy to believe, always provided that one could perform the feat of convincing oneself that the earth, apparently the very pattern of immobility, was none the less moving more swiftly than anything known to mankind. Next Galileo's telescope revealed some appearances that agreed well with the Copernican view, but not with the older systems. Then Kepler, using the Copernican system, but with elliptic orbits, gave the first exact laws of planetary motion. Newton crowned all this by showing that this system and Kepler's laws could be mathematically explained, if it were assumed that heavenly bodies obeyed the same laws of motion as earthly bodies, and that all bodies attracted each other in proportion to the product of their masses and inversely as the square of their distances. To do this Newton had first to show how the ordinary bodies on earth moved, that is to say, he had to discover a system of mechanics. Others, it is true, had made a beginning, but it was Newton who first defined such ideas as mass, momentum, force, inertia and so forth, and set out dynamics, like geometry, in a series of inescapable propositions. He could not experimentally prove the attraction between earthly bodies, nor could anybody do so for a century after, but he did at least show that the observed motion of the moon could be explained by assuming such an attraction. The title of his great book *Mathematical Principles of Natural Philosophy* tells us what he did. It was not his observations in astronomy that were important, but his success in showing that all motions, whether on earth or in the heavens, were expressed by the same laws. This is not to say that Newton was an armchair scientist, for his work on light (p. 74) is enough to witness the contrary. It took the world a little time to take in what Newton had done, but after about 1730 everybody enthusiastically hailed his picture of the universe as the true one, and set to work to deduce its consequences.

The study of living organisms had not by the year 1600 advanced far beyond Aristotle, except in respect of human anatomy; but one great new field was soon to be opened—the realm of things too small for human sight. Francis Bacon, himself no experimentalist, did much in these years to persuade men of the powers of experiment, and emphasized his conviction that in all things there was a fine structure, as yet unknown, the knowledge of which would explain many of their properties—a profound truth which has fairly come into its own in this the atomic age.

Soon after he had made his telescope, Galileo found out that by using a pair of lenses of the right focal length he could obtain an instrument that would magnify small objects—what we now call a microscope. This invention he did not much develop and it seems to have hung fire for quite a time. To make the small strongly curved lenses for a good microscope was much harder than to make the long-focus lenses for a good telescope, and little microscopical work was done before Robert Hooke redesigned the instrument and taught the instrument makers the necessary technique. Hooke has been called the first professional scientist. For most of his life he was 'curator of

Fig. 51. Robert Hooke prepares a flea for inspection with his microscope (*c.* 1664).

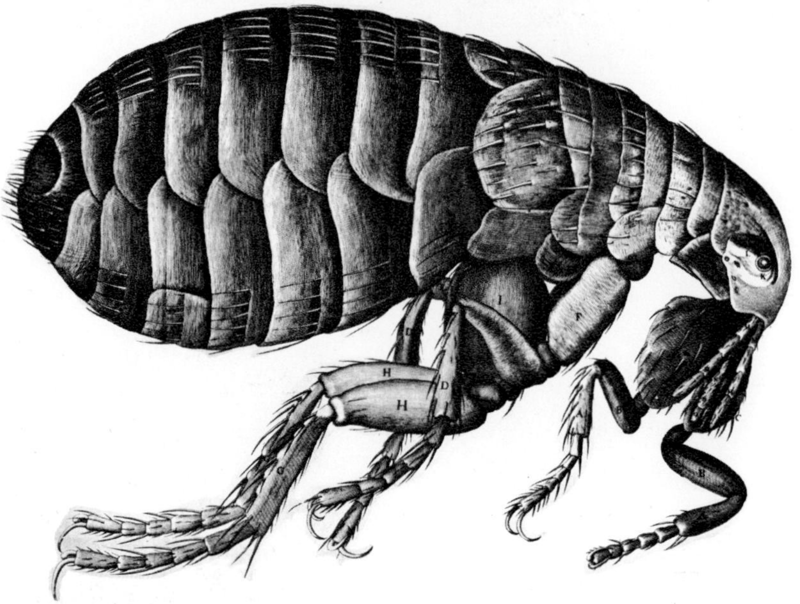

Fig. 52. A much-reduced reproduction of Hooke's engraving of a flea, from his *Micrographia*, 1664.

experiments' to the Royal Society, and before each meeting prepared the experiments which the Fellows wished to see performed. He was, as befitted the post, a master of experimental technique. The work gave him the opportunity for research, and a great deal of this appears in his chief publication, *Micrographia* (1664). This book contains the account of his microscope and of many of the objects he observed with it: it is full of profound and fruitful scientific ideas, some of which have had to wait till modern times for development. But though Hooke showed how to make and use the microscope he did not examine living creatures in any systematic way. The same is true of another great microscopist, the Dutchman, Antony van Leeuwenhoek, who made microscopes of powers much greater than ever before. Leeuwenhoek performed the remarkable feat of grinding very minute and nearly spherical glass lenses and of mounting them between two metal plates. Now the greater the curvature of a magnifying glass, the greater the magnification; so it is easily understood that a tiny seed-like sphere of glass can give a very large magnification. Leeuwenhoek's microscopes were by no means easy to use and the chief difficulty was to mount the object so that anything could be seen. The field of vision was minute and was sharp only at the very centre. Leeuwenhoek's skill and patience, however, enabled him to discover very many minute forms of life, including even bacteria, and the consequence of his work was the discovery of a totally unexpected world of life too small to be seen by

Fig. 53. Antony van Leeuwenhoek uses his small but effective single-lens microscope (*c*. 1680).

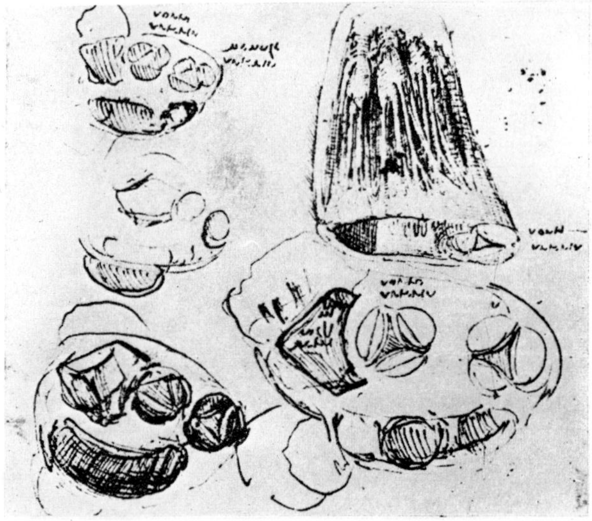
Fig. 54. Valves of the heart as drawn by Leonardo da Vinci.

the naked eye. Leeuwenhoek's most famous discovery was that of spermatozoa, thus throwing a wholly new light on the generation of animals.

The microscope soon began to impress on man the significance of the fine-structure of things. Thus Aristotle supposed 'flesh' to be a continuous *stuff* like butter or glass, but even the imperfect microscopes of the seventeenth century showed that it was a most complex structure of fibres, blood-vessels, etc. Men began to think of their bodies as mechanisms, rather than so much stuff controlled and operated by soul and spirit.

This discovery is linked with one of the very greatest of the age, that of the circulation of the blood. The ancients knew that the heart moved the blood, but supposed that its motion was a simple ebb and flow. How was it that these men, who had dissected human and animal bodies with some care, failed to realize that the blood circulated round the body, passing from the heart outwards through the arteries, then through the organs and back through the veins to the heart? The chief and sufficient reason was that there did not seem to be any passages by which the blood could pass through the organs. The arteries and veins branched into smaller and smaller twigs which disappeared in the fleshy mass, but no blood-vessels could be seen to pass through it. We now know that there are microscopic vessels—capillaries—that pass from artery to vein, but without a microscope they could not be seen. Yet if they could not be seen they could be inferred and William Harvey made his great discovery of the circulation of the blood about 1616 without seeing the vessels it traversed. Harvey came to his conclusions in the true seventeenth-century way: he observed and he measured. He watched the isolated heart of a tortoise beating outside its body: he realized that the valves in the veins and heart, which others had seen but not understood, ensured that the blood could move only in one direction and could not ebb and flow as the ancients supposed. Then he reckoned from the capacity of the heart how much blood went through it at each beat, and he found that in an hour more blood passed through it than

Fig. 55. William Harvey prepares for a dissection.

Fig. 56. The frontispiece of a sixteenth-century edition of Witelo's book on optics, illustrating natural phenomena of reflection and refraction.

was contained in the whole body! Arguments such as these convinced him that the blood circulated, but he never saw the passages by which it oozed through the muscles and other organs. This left a little room for doubt, but half a century or so later Malpighi and Leeuwenhoek actually saw through their microscopes the blood travelling in the capillaries: and the world was finally convinced.

A very great advance towards the understanding of the world resulted from Newton's studies of light. Before his time the study of light was mainly a problem of geometry, and therefore acceptable to the Greeks and Arabs. Fig. 56 is the frontispiece from a sixteenth-century edition of the best mediaeval book on light, and it gives you a very good idea of the problems studied in the old optics. The transmission of light—reflection and refraction—was a problem in geometry, but the discovery of what light was seemed to be a philosophical problem rather than a scientific one. In the seventeenth century, this problem

Fig. 57. Newton splits up sunlight into the spectral colours by means of a prism.

had become very interesting and widely differing views were held. Was light a quasi-spiritual influence, a sort of force, a stream of particles, or a vibration—as sound was known to be? What was it that made the difference between, let us say, red light and blue? Had light a measurable velocity or was it transmitted instantaneously?

It had long ago been noticed that rainbow colours could be produced from white light: Roger Bacon in the thirteenth century and Alhazen even earlier tell us how waterdrops and hexagonal crystals in sunlight show these colours, but Newton was the first to study this experimentally. He solved a great part of the problem when he was twenty-three, the age at which he was most intensely interested in scientific matters. In one year he discovered the binomial theorem, the differential and integral calculus, the theory of gravitation, and the theory of colours, though, of course, these great ideas took many years to develop. Newton became interested in rainbow colours because he found that his telescope lenses always gave images fringed with colour and to that extent indistinct. So he bought himself a prism which, as he knew, showed these rainbow colours better than anything else. He bored a hole in the shutter to let a ray of the sun into his darkened room, and put his prism in the path of the ray so that the light should be refracted on to the opposite wall. He then saw that instead of a circular spot of light a long strip of rainbow colours was seen on the wall: moreover he isolated light of each of these single colours by screening off the others and showed that it could not further be broken up by the prism, whence he concluded that, unlike white light, these colours were not mixed but 'homogeneal'. He therefore concluded that the white light of the sun was in fact a mixture of all these colours and that the different coloured lights in that white light were bent by the prism to different extents and so were thrown upon different parts of the wall.

Fig. 58. Newton's rings.

This was Newton's first publication and it raised so much controversy that he nearly gave up science altogether. But he went on to think out what light might be. He could not believe it was a wave-motion like sound, because sound travels round corners and light does not seem to do so: and he favoured the view that light was a stream of very minute particles travelling at a very high speed. Nevertheless, he thought there must be

Fig. 59. Newton with his original reflecting telescope, still in the possession of the Royal Society.

something like an oscillation in it, in order to explain the colours of films, which Hooke had already explained in a fashion by a wave-theory. Very striking was Newton's demonstration of the succession of dark and light rings which are formed when a convex lens is pressed against a flat piece of glass. The dark and light lines alternate, but there is nothing that alternates in the lens, the glass or the air between them, nor yet in a stream of particles of light, whereas if light consisted of waves, their succession could provide the source of alternation. Newton's views were not settled, but he put forward the theory that light travelled as particles in straight lines, but was capable, on hitting matter, of setting up vibrations in the ambient ether. About the same time another very great scientist, Christian Huygens in Holland, developed a wave-theory of light, but thought of it as longitudinal waves like sound, whereas its waves are actually transverse, like those of the sea.

The world was inclined to follow Newton, and from his time until the years following 1800 light was regarded as consisting of particles. From 1800 to 1900 it was thought of as waves in the ether: from 1900 on we have come to think of it as having properties both of particles and of waves—a view not much unlike that of Newton.

One important practical invention followed from the new theory of colours. Newton supposed that, since the various coloured lights that made up white light were refracted to different extents, lenses, which are refractors of light, must necessarily give images made indistinct by colour-fringes; he therefore tried to replace lenses by curved mirrors, which reflect all colours in the same direction. To this end he invented the reflecting telescope, which was gradually perfected in the next century. Later it was found that, by using combinations of lenses made of different glasses, these colour fringes could be so far reduced as to cause little inconvenience, and these so-called *achromatic* lenses were used for telescopes after 1760 and for microscopes after 1825.

The defects of the lenses of the time made the early microscopes (other than some of Leeuwenhoek's) incapable of magnification of any high order, and consequently the discovery of most of the very fine details of living things had to await the better instruments of the nineteenth century. But the microscopes of the years 1670–1820 gave quite good images magnified up to 100 diameters or less and there was an enormous amount to be discovered about the objects and structures so revealed. In this period the anatomy of insects and the structure and classification of the minute living forms familiar to the student of pond-life were well worked out, and some amazingly good work was done by men of skill and patience, using the crude instruments of the age. Swammerdam's dissections of insects are a shining example.

In this glance—for it is no more—at the brilliant and fundamental discoveries of the seventeenth century, I have omitted all reference to the sciences of chemistry and electricity. They were indeed studied at that time, but the fundamental discoveries came only in the eighteenth century, and I have therefore postponed their consideration till my next chapter, which deals with the science of that age.

Why did the seventeenth-century scientists achieve so much more than any before them? First and foremost, because they were not content to sit at their desks and make books from other books, but set up their laboratories and saw for themselves what happened. Moreover they did not merely look at what happened but they measured and weighed and timed it. They wrote down what they did and discussed it in their societies and published it in their books and journals. Indeed they did very much what scientists do today. It is true that they lacked our advantages. There were very few scientists and fewer paid jobs for them. Their instruments were crude and relatively costly; their laboratories were without gas, electricity or piped water. The scientist of the beginning of the century had to make most of his apparatus, or at least supervise the men who made it. This put many scientists at a great disadvantage, for although some, like Galileo, were natural craftsmen and rejoiced in mechanic's work, others, though fertile of ideas, were unskilled with their

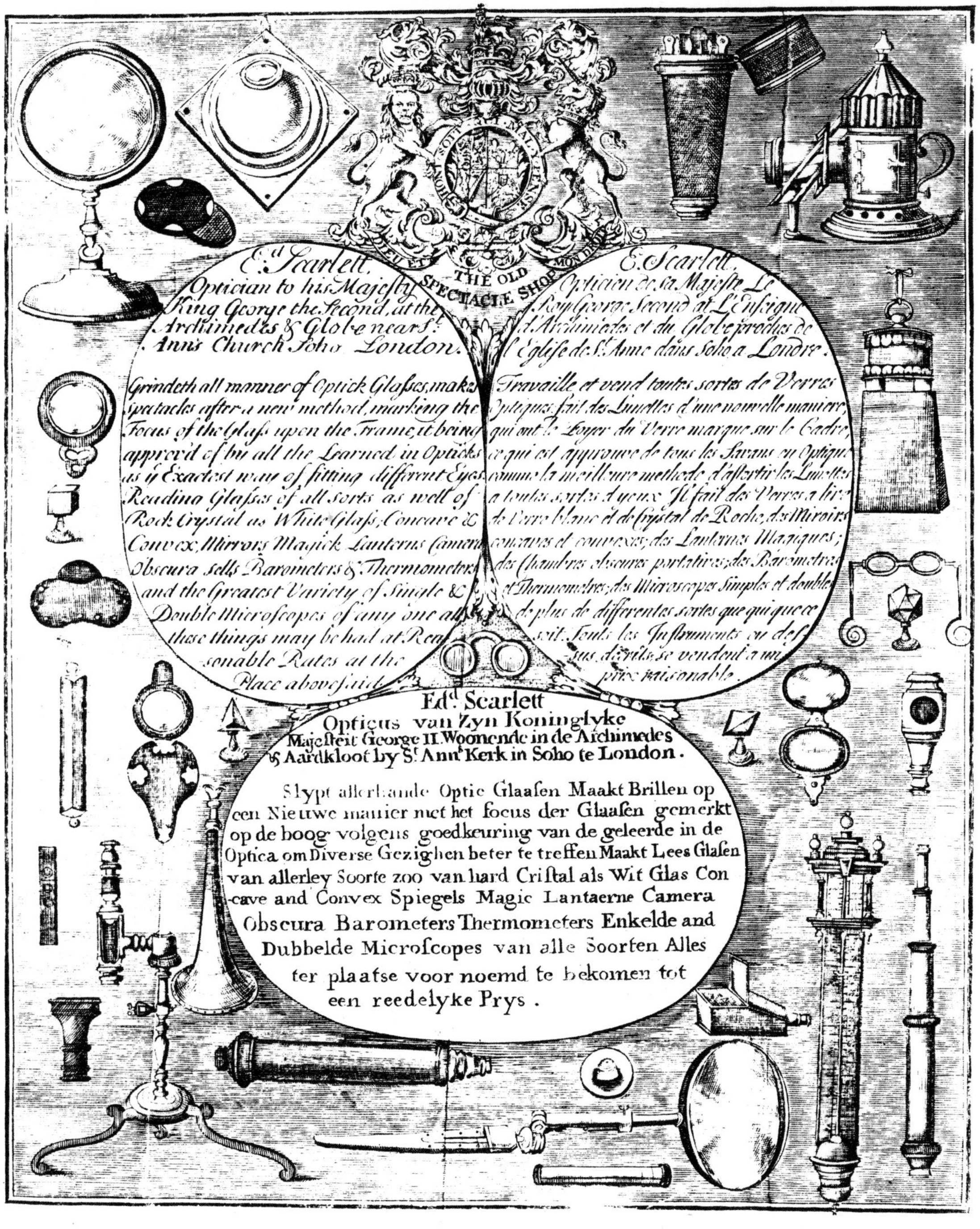

Fig. 60. A trade-card of the early eighteenth century, showing the instruments sold by Edward Scarlett.

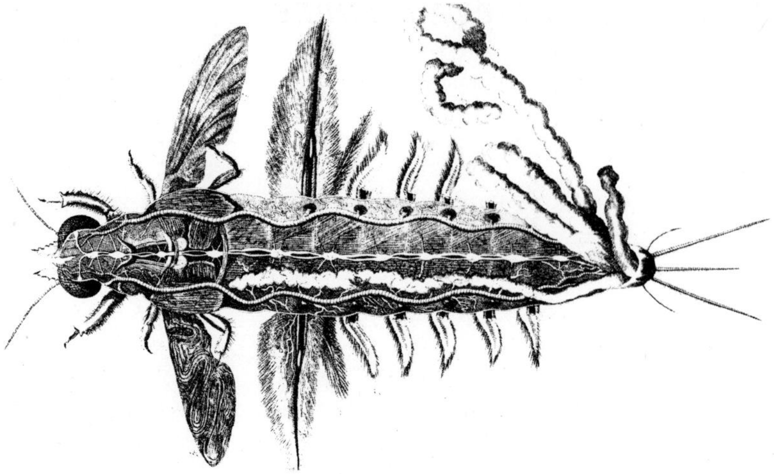

Fig. 61. A triumph of skill. Swammerdam's dissection of a May-fly.

hands: consequently the growth of the trade of instrument maker went far to ease and quicken the progress of science. In the year 1600 and even before there were a few instrument makers, but almost entirely in the astronomical and mathematical field. They would supply to order beautiful astrolabes, sundials or drawing instruments, and as new types of scientific instruments came to be demanded by the general public these men learned to make them. The popularizing of clocks in the latter half of the century, after the adoption of the pendulum had made them reliable, created a great demand for craftsmen of a high order of skill; for the man who could make a clock could make most other instruments. By the end of the century, the instrument maker's was an important trade, and anyone who could afford to do so could buy a microscope, a telescope or an air-pump. It thus became easy to acquire a laboratory, and in this respect the scientists of the eighteenth century found their paths made smooth.

CHAPTER THREE

SCIENCE IN THE EIGHTEENTH CENTURY

BY THE YEAR 1700, SCIENCE was not only recognized but respected. It began indeed to be thought of as a kind of knowledge that a man who wished to understand the world ought not to be without. So, although it was long before science became a regular part of the education of any but medical men, courses of lectures began to be provided in universities and in such cities as could provide an interested audience. In the eighteenth century every civilized country had its scientific society, of which our Royal Society was the most famous. All the sciences continued to receive attention, but two of them made progress which transformed them into major departments of scientific endeavour. These were electricity, with which magnetism may be coupled, and chemistry, which passed from the alchemical to the scientific.

The study of electricity and magnetism began long before its time of rapid progress. As we have seen, the mariner's compass, first heard of about 1200, had become an important means of navigation in the fifteenth century; and naturally men began to ask why the loadstone or magnetized iron pointed to the north. The important question seemed to be, '*To what real thing does the compass point?*' To the mediaeval mind the natural answer was the pole-star, for the stars were the great exerters of influences on every kind of body. But very soon (at least as early as 1436) mariners found that the compass did not point to the true north (as determined by observation of the sun or the pole-star) but a few degrees away from it. It was then commonly thought that the compass pointed, not to the pole-star, but to some 'invisible point'. Columbus found that the variation of the compass (the angle between its needle and the true north) steadily altered as he proceeded westwards across the Atlantic. In 1544 a new phenomenon was noticed, that a compass needle, pivoted on a point so that it could move vertically, did not point along the earth's surface, but down into the earth: in 1576 Robert Norman, a London maker of mariners' compasses, measured the angle of dip with care, but did not understand its significance. But in 1600 a great English scientist, William Gilbert, published in latin a book on the magnet in which he summed up a life's work. He began by confuting a pack of old wives' tales about magnets, and then introduced his great idea—*that the earth itself was a magnet.* Unlike so many of the scientific writers of earlier times, he was not content with an idea, but went on to test it by experiment. To this end, he constructed a model earth that was a magnet, by making a small globe out of a piece of loadstone. This he called a terrella

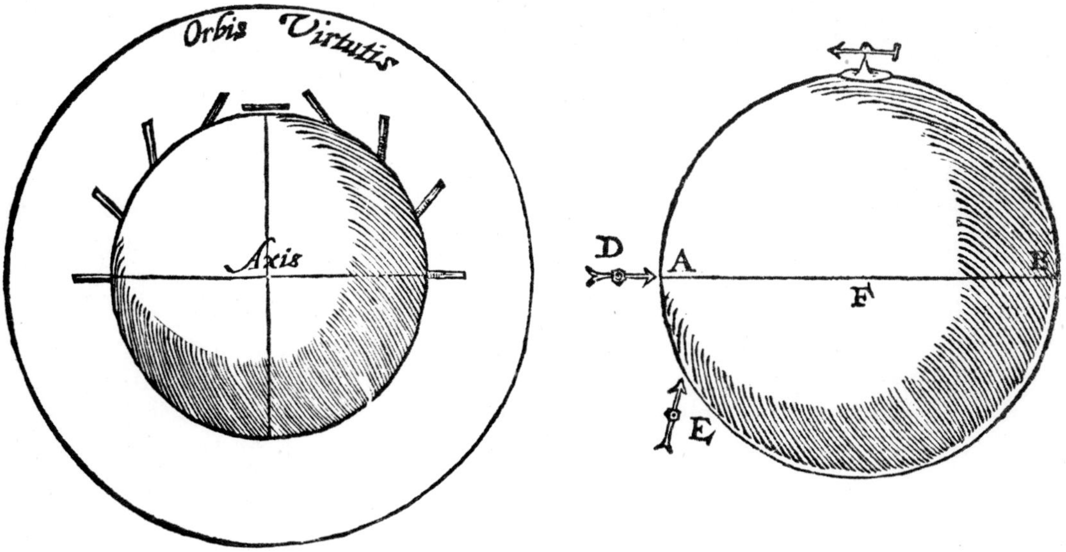

Fig. 62. Gilbert's terrellae, with the needles upon it pointing to the poles and showing the angle of dip.

or 'earthlet'. Gilbert's most important experiments with his terrella were to place upon it a series of little iron wires or small compass needles, whereupon they behaved just as compass needles do on earth, pointing along the meridian and showing the expected dip, thus showing that the earth affected compasses in the same way as a globular magnet.

Gilbert set down most of the elementary facts about magnets without serious error, but his enthusiasm for magnetism carried him a little too far, for he supposed that it was the force that operated the movements of the earth and planets about the sun—not an unreasonable idea, since magnetism and electricity were in his time the only proven examples of attraction. Terrellas are great rarities today, but two have been preserved by the Royal Society, which has lent them to the Science Museum. They were in the Royal Society's museum in 1681 and I suspect they are the loadstones about which this verse was written in 1660 or 1661.

These men take nothinge upon trust
Therefore in Counsell sitte many howres
About filing iron into dust;
To experiment the Loadstones powers,
In a circle on a board they strew it
By what lines to see the loadstone drew it.

Fig. 63. Fellows of the Royal Society studying the lines of force of the loadstone. *Back row:* Dr. Wilkins, Sir Robert Moray, John Evelyn. *Front:* Lord Brouncker, Robert Boyle, Sir Christopher Wren.

Fig. 64. William Gilbert's electroscope.

But despite this interest in the magnet no great progress beyond Gilbert was made for many years.

Let us now return to electricity. We have seen that in ancient times men knew that amber (in Greek, *electron*) when rubbed attracted light bodies, but the idea of electricity and the word itself dates from the sixteen-forties. The first scientific study of electrical attraction was made by William Gilbert before 1600. He made the first electroscope, a pivoted metal needle (Fig. 64), and by its aid he discovered that not only amber and jet, but many other materials such as rock-crystal, glass, most precious stones, mica, sulphur, rosin, mastic and sealing-wax attracted it. This attraction, he proved, was not a special property of amber, but was common to many substances.

No great advance on this was made until, sixty years later, von Guericke, inventor of the air-pump, made the first electric machine. It was a ball of sulphur mounted on a wooden axle. When rotated and rubbed, it attracted light bodies and then repelled them, and the inventor of it discovered that this power could be transmitted through a short string. He thought of this globe as an explanation of the attractive power of the earth, rather than as a means of making electricity. Robert Boyle and Isaac Newton were both puzzled about this attraction. Newton demonstrated that if he rubbed the top of a glass plate bits of paper beneath it were attracted. This seemed to indicate the generation of something that travelled through the glass; he thought of it as some kind of very minute and sticky particle.

With the beginning of the eighteenth century (1709) we find Hauksbee in England discovering the effectiveness of glass as a means of producing electricity. He made an electrical machine consisting of a glass globe rotated mechanically against a rubber of coarse woollen cloth. In this way he produced some remarkable luminous effects. Interest was aroused by this curious phenomenon, which was really a very early forerunner of the modern discharge-tubes, but it remained an unexplained and isolated fact. It may be said that, up to the end of the first quarter of the eighteenth century, scientists had found electricity an odd unaccountable phenomenon, and had not found out any of the laws that describe it.

In 1729 an Englishman, Stephen Gray, discovered the property of electricity which makes possible all our electrical devices, namely that it can be conducted, apparently instantaneously, for long distances. The very high-tension electricity which is made by friction is readily conducted through even poor conductors and Gray found that sticks and thread would carry electricity from a

Fig. 65. Otto von Guericke demonstrates the attraction of light bodies by means of his electrified sulphur globe.

rubbed glass tube to an ivory ball, which would then attract a feather, or a piece of gold leaf. In the course of these experiments he found that if he supported the conducting threads with other threads or wires he got no effect (the electricity being, in fact, conducted to earth), but that if he used silk to support it, the electricity passed along it to the end; thus he arrived at the notion that some materials conducted electricity and some did not. In 1734 he concluded that the best conductors were metals, and so brought into use the electric wire.

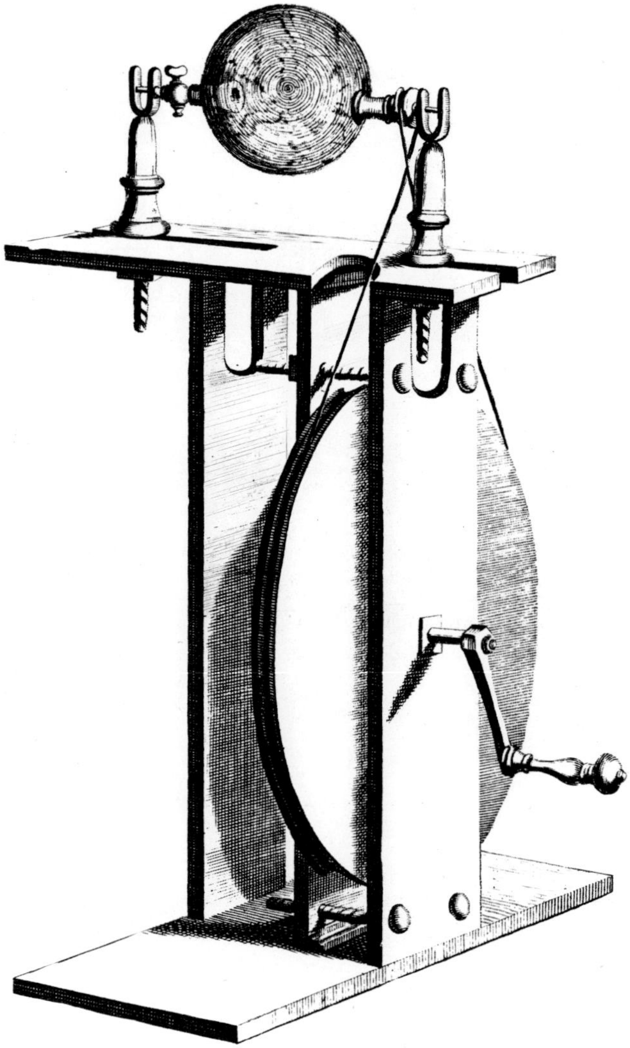

Fig. 66. Francis Hauksbee's Electric Machine.

Fig. 67. Stephen Gray (*left*), with the assistance of Granvil Wheler, demonstrates, in the latter's house at Otterden Place, the conduction of electricity through a long thread, and at the same time discovers the distinction between conductors and non-conductors.

Next came the discovery that there are two kinds of electricity. A botanist, Charles Francois de Cisternay Du Fay in France, about 1733, noticed that a piece of leaf gold, electrified by means of a rubbed glass tube, was repelled by the glass, but attracted by rubbed gum copal. Evidently the electricity of glass was different from that of copal. He made a simple piece of apparatus (Fig. 68) to test the matter further. He balanced a piece of rubbed amber on a pivoted wooden ruler and he found, that on bringing various rubbed substances near to it, that some attracted the amber and some repelled it. The kind of electricity that crystal and glass gave he called 'vitreous'; the kind that amber and resins gave, he called 'resinous'.

These important experiments were done with rubbed glass tubes and the like, but after 1740 the Germans revived Hauksbee's electric machine as a source of electricity and many devices for rubbing glass cylinders with silk or leather and collecting the electricity from them were made. With these frictional machines much more powerful effects were attained. Thus in 1744 it was shown that electric sparks were hot, for they were able to set ether alight. In consequence of such experiments electricity came to be thought of as something like the element of fire, in which the world still firmly believed, and was sometimes alluded to as the 'electric fire'. The frictional machine yielded only very minute quantities of electricity and much greater effects were obtained after it became possible to accumulate the electricity of the frictional machines in the Leyden jar, discovered independently in Germany and Holland in 1745. The first form it took was a bottle containing water (or alcohol or mercury), with a long nail dipped in the liquid and projecting from the neck. The discoverer held the flask in his hand and connected the nail to an electrical machine, presumably with the intention of 'electrifying' the water; but on touching the nail he was astonished and alarmed to feel an electric shock—perhaps never before felt by man, except from lightning or the electric eel, the shocks of which were not at first realized to be the same as that of the scientists' electricity.

It was the electric shock that put electricity in the news. Everyone wanted to feel it—or see someone else suffer from it—and lectures on electricity became very popular. At first the effect of the Leyden jar was thought to be due to electricity stored in the water. But in attempts to improve the jar, the outside was coated with tinfoil, and soon afterwards Dr. Watson coated the inside with tinfoil and dispensed with the water.

The Leyden jar gave great opportunities for spectacular experiments, of which the most bizarre was to pass the shock through some hundreds of Carthusian monks, joined by iron wires; all gave a simultaneous jump when the jar was discharged! It was now possible to try to discover how far electricity could be transmitted and whether it was transmitted instantaneously. Before

Fig. 68. Du Fay performs the experiment which showed that an electrified body is attracted by some electrified bodies but repelled by others, and is thus led to suppose two kinds of electricity.

Fig. 69. The Abbé Nollet was one of the first to feel the shock from a Leyden jar.

1750 it proved possible to send a shock through two miles of wire, nor could any interval between the discharge of the jar and the sensation of the shock be detected.

We now come to the discoveries of that great American, Benjamin Franklin. In the years round 1750 he cast doubt on the existence of two different kinds of electricity. Two people stood on cakes of wax to insulate them. One rubbed a glass tube and let a charge from it pass to the other. Both now appeared to be oppositely electrified, but there had been only one kind of electricity present. If now they touched hands a charge passed and the electricity disappeared. Where had it gone to? Franklin thought that the two kinds of electricity (called by Du Fay 'vitreous' and 'resinous') were produced by one single electric fluid. A body charged with vitreous electricity contained an excess of electric fluid and was said to be positively electrified: a body charged with resinous electricity had a deficit of electric fluid and was said to be negatively electrified. Charging a Leyden jar was simply transferring electric fluid from one coating to the other. Finally Franklin proved that the charge was in the glass and not in the coatings.

These ideas were very fruitful for electrical science, but Franklin first astonished the world by his proof that lightning was no more than a very large electric spark. Others had suggested this, but Franklin, after a careful comparison of lightning and electricity, thought of a way to prove it by experiment. He prepared a kite, armed with a pointed wire, and sent it up just when a thunder-cloud was passing. The loose threads of the kite-string stood out and

Fig. 70. The Abbé Nollet's demonstration that the conduction of electricity appeared to be instantaneous.

Fig. 71. Benjamin Franklin draws electricity from a thunder-cloud.

soon he drew a spark from a key he had hung on the string. Then the rain came down and wetted the string, making it a better conductor: electricity then passed freely down the string and was shown to do all that electricity normally did. The discovery was not only interesting but useful, for it led to the invention of the lightning conductor, still in use today.

The most powerful means of investigating any scientific subject is measurement: but up to the seventeen-sixties no electrical measurements whatever were made. So we must celebrate two great men, Henry Cavendish and Charles Augustin de Coulomb, who were pioneers in this field. Cavendish, an eccentric solitary millionaire, tried to measure the relative conducting power of solids, liquids, solutions, etc. To do this he probably used himself as his own galvanometer, and estimated the currents which passed through these conductors from his Leyden jars by the shock they gave him! He also investigated the capacity of condensers and proved the inverse square law for electric charges: but most of this wonderful work, which was twenty-five to fifty years ahead of his time, he left unpublished.

The clear establishment of the fact that electrical attractions and repulsions follow the same type of law as gravitational attraction—that they are proportional to the product of the charges and the square of the distance between them—was a tricky thing to prove because it involves the measurement of very small forces. Coulomb discovered that the angle through which a wire or other filament is twisted is a measure of the force twisting it. He hung, by a filament of silk, a horizontal rod with a ball at one end (Fig. 72). To this ball he gave an electrical charge, and brought up another charged ball towards it. The charged balls repelled each other, and the force with which they did so was finally balanced by the twist of the silk. He measured the angle through which he had to twist the silk in order to bring the balls nearer to each other by a given distance and so worked out the way the force of repulsion between the balls depended on the distance between them.

Fig. 72. Coulomb's torsion-balance.

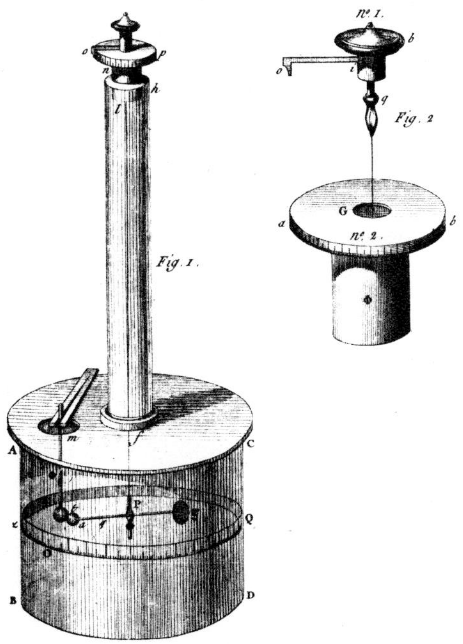

This work marks the limit of the researches that the scientists of the eighteenth century could make with the minute high-tension currents obtainable

Fig. 73. Robert Hooke assists Robert Boyle in investigating the effect of air on respiration:
'And out of the glass the air being screwed,
Puss died and nothing so much as mewed.'

from frictional machines. A new chapter of electrical history opened with the discovery of the electric battery and its large low-tension currents, but this belongs to the nineteenth century and our next chapter.

If an intelligent seventeenth-century man-in-the-street—someone like Samuel Pepys—had been asked what chemistry was, he would very likely have replied that it was the art of separating and compounding substances so as to make new medicines. If a Victorian or even a modern schoolboy were asked the same question they might have replied, 'It's about gases.' The development of chemistry from an art to a science came about in the eighteenth century through the study of gases—or, as they said in those days, 'different kinds of air'.

The origin of this interest sprang from the air-pump of von Guericke. This was redesigned and much improved about 1660 by Robert Boyle and his assistant Hooke: consequently it became not only possible but easy to obtain a space free from air, and so discover what phenomena would not take place without air. These experiments were the beginning of a long struggle to discover what air had to do with burning and breathing. Robert Boyle used his air-pump to show that substances which would burn in air would not burn in a vacuum. He put a cylinder of hot iron in his receiver, pumped out the air and, by a mechanical device, dropped on to the iron some sulphur, a substance that in air catches alight at a relatively low temperature. It did not catch alight in the vacuum. But he knew that gunpowder would burn in a closed gun-barrel: so he tried the same experiment with gunpowder: it did not explode in the vacuum but burned with a bluish flame. So it seemed that either air or else something that was in gunpowder—i.e. saltpetre—was needed for combustion. So it looked as if something necessary for combustion was contained both in air and in saltpetre.

Animals placed in the receiver died when the air was evacuated and so air seemed to be necessary for life as well as for combustion. But what part did the air play? Did it only carry away waste products, or was it actually used up by burning and respiration? The question was answered by the experiments of Robert Mayow in 1674. He showed that when things burned or animals breathed, a part of the air disappeared and a part was left, and that this latter part was no longer able to be used for burning or breathing.

So Mayow supposed the air had in it a 'nitro-aerial spirit' that was used up in combustion and respiration. The idea of a 'spirit' was nearly the same as that of a gas, so Mayow came very near to the idea of oxygen. Robert Hooke advanced similar ideas, but they met with great resistance: first, I think, because people could not bring themselves to believe that air was not an element —a simple substance—as had been believed since ancient times, and secondly, because the universal and age-long view of combustion was that some fiery

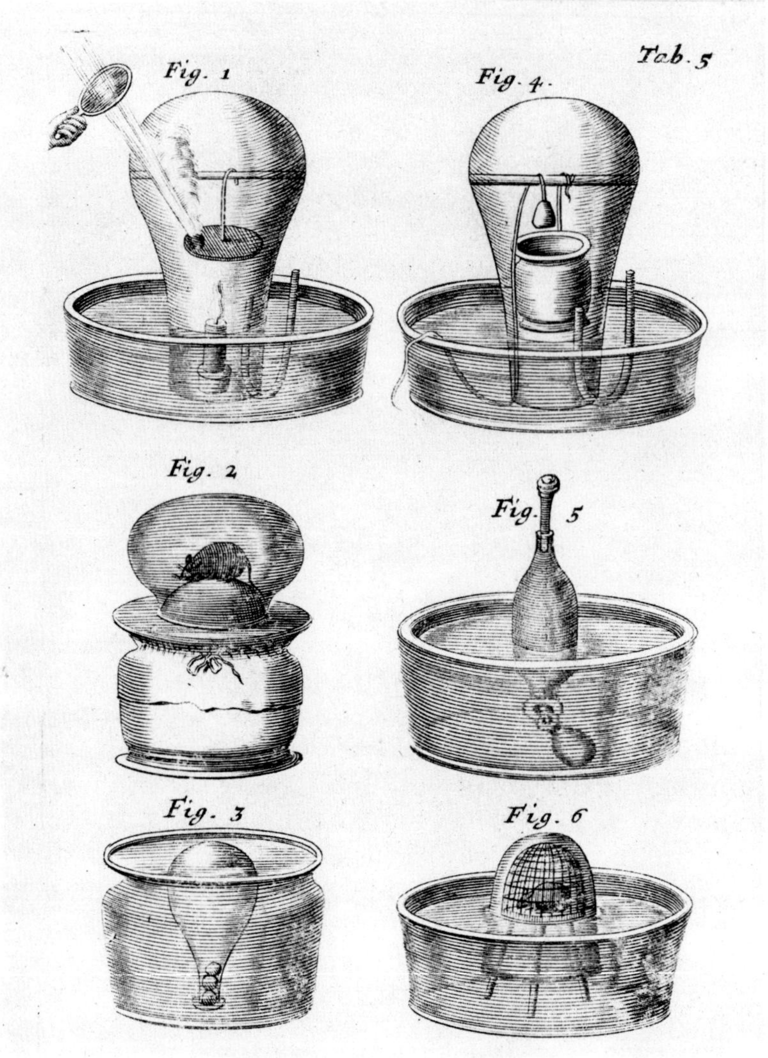

Fig. 74. The apparatus by which John Mayow demonstrated that only a part of the air was concerned in respiration and combustion.

substance came out of the burning body, as indeed appeared to the eye. It seemed absurd to say that air went into it. Yet there was already some evidence for this. Several people had noticed that when metals were converted into 'calx' (we would say 'oxide') by heating them they became heavier. Boyle tried to find out whether this was the result of air combining with the metal, but his experiments were badly planned and led to nothing. The world then lost interest in these problems for a century and went back to the ancient

Fig. 75. A seventeenth-century pharmacist calcining antimony to its calx (oxide) by means of a burning-glass.

theory that things which would burn contained a sort of fiery matter (phlogiston), which issued from them when they burned.

Another approach was needed, and this came through the technique of isolating, handling and recognizing the different gases. J. B. van Helmont before 1640 and Robert Boyle a little later knew that there were several different kinds of gas or air, but they did not work out a technique of handling them. Complementary to their work was that of an English clergyman, Stephen Hales, who in the seventeen-twenties found out how to handle and measure gases, though he did not recognize that there were different kinds! Hales was a great physiologist who did pioneer work on the pressure of sap and blood, but his contribution to chemistry was the discovery that a great many things when heated emitted what he called 'air', the quantity of which he measured. He was the inventor of the technique of collecting gases over water.

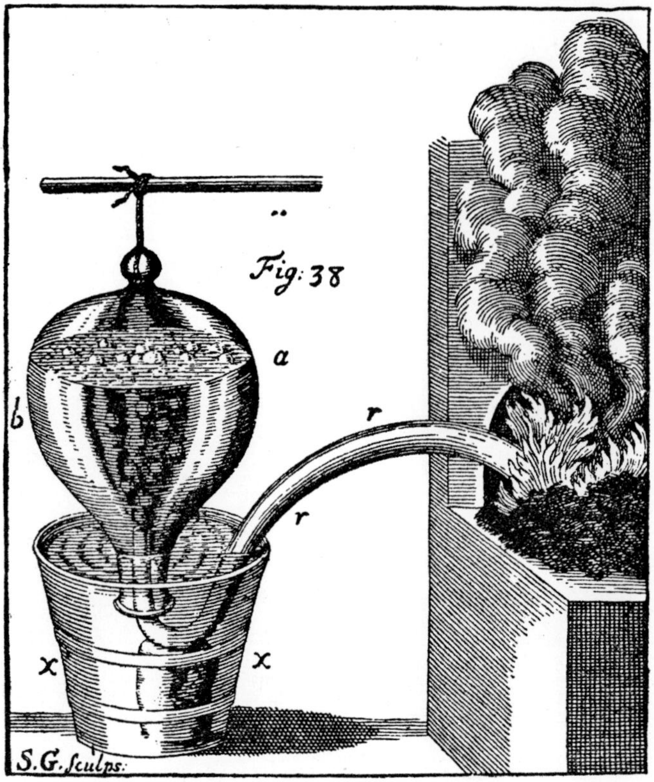

Fig. 76. How Stephen Hales collected the 'air' given off when substances were heated.

In the course of his experiments he did in fact make several gases then unknown, such as oxygen, methane and coal-gas, but he let them escape without considering whether they differed from ordinary air; thus he did not really study gases in a chemical way at all. There was no further progress until 1754 when Joseph Black investigated the relationship between chalk and lime, and the different kinds of magnesia. He found that chalk (or limestone) was quicklime combined with what he called 'fixed air'. He collected and studied this gas, which we now call carbon dioxide, and for the first time he removed all doubt that there were kinds of air as different and as well-defined as were lime or alum or copper or any other of the liquid and solid substances then known to the chemist.

Twelve years later Henry Cavendish made inflammable air (which we call hydrogen). He thought of it as a kind of air, but with a great deal of the 'fiery matter'—phlogiston—in it.

In the seventeen-seventies, Karl Wilhelm Scheele in Sweden and Joseph

Fig. 77. Henry Cavendish as a young man, with the apparatus which he used for handling gases and with which he isolated inflammable air (hydrogen).

Priestley in England independently took up the investigation of gases and between them discovered at least seven new ones. Both of them independently discovered what we call oxygen. Scheele made it from nitric acid, and several other sources and called it 'fire-air': Priestley made it from mercuric oxide and called it 'dephlogisticated air' (air which contained little or no phlogiston and would draw phlogiston out of other bodies and so make them burn). Scheele was the first to make oxygen and recognize it as a new gas, but Priestley's discovery was the first to be published.

Here was a new kind of air possessing the activity of common air in a higher degree. Bodies would burn in it more brightly and animals would survive longer in it than in the same measure of common air. In 1772, a little before these experiments, Daniel Rutherford had examined the residue of gas left behind after animals had breathed or charcoal had been burned in air: from this residue he absorbed the fixed air (carbon dioxide) by means of caustic potash and found that a new gas unaffected by alkalis was left. This was the gas we now call nitrogen. It would seem very easy to put these experiments together and reach the idea that air was a mixture of oxygen and nitrogen, and indeed Scheele had done so, though he had not published his views. But Priestley's head was full of the notions of phlogiston and the elementary nature of air, and it was left to Antoine Laurent Lavoisier to clear up the whole question of the nature of air, water, combustion and respiration.

Just about the time when Scheele and Priestley discovered oxygen, Lavoisier repeated Boyle's experiments in a better way and so proved that when lead and tin were heated in air contained in sealed vessels they took up some of the air and increased in weight. He burned phosphorus in air confined over mercury under a bell-jar, and he noted that the air diminished in volume. He collected the white solid residue, which we now call phosphorus pentoxide, and weighed it: it was heavier than the phosphorus. Clearly when some things burned, air combined with them, but Lavoisier had not yet appreciated that it was only a part of the air that did so. Then he heard of Priestley's discovery of 'dephlogisticated air', hit on the answer to the problem of air and combustion and went out to prove it. To do so he devised the beautiful and famous experiment illustrated in Fig. 80. In a retort (or drawn-out flask) he put four ounces of mercury and introduced the neck of the retort into some air confined over mercury in a bell-jar. The retort, tube and jar contained 50 cubic inches of air. By means of a small furnace fed with red-hot charcoal he kept the mercury just below its boiling-point for several days on end—a very skilled and tricky business. Gradually red specks and scales of mercuric oxide formed; then, after twelve days, the fire was let out and the experiment stopped. When all was cold there were 42 cubic inches of gas left. This was shown to be 'mephitic air' (the gas that was later called nitrogen). Then Lavoisier

Fig. 78. Scheele prepares oxygen by heating nitric acid and allowing the fumes to pass into a bladder containing some lime and water.

Fig. 79. Cavendish's eudiometer.

collected the red scales of mercuric oxide and heated them in a tiny retort, whereupon they gave off 8 cubic inches of 'pure air' or 'eminently respirable air' which he later called oxygen. This was just as much as the original air had lost. Finally Lavoisier mixed the nitrogen and oxygen and the result was common air.

So Lavoisier asserted that substances burn only in this 'pure air' (oxygen), that when they do so they combine with it and that phlogiston was only a name. As time went on he accumulated more and more facts contrary to the theory of phlogiston, the fiery element, and in the course of the next twenty years the idea was generally abandoned. Lavoisier now founded a new sort of chemistry. He would have nothing to do with the old 'elements' of earth, air, fire and water, or the 'principles' of salt, sulphur and mercury. He followed Boyle and said, 'We will call a body an element if we cannot decompose it.' He made a list of some twenty-five such elements, and in effect said, 'Here is a list of elements you can weigh, measure and observe: we will explain the composition of other bodies in terms of these.' Thus he relieved himself of the prejudice that phlogiston must exist and that air and water must be elements, and was in a strong position to solve the remaining great problem of the composition of water.

Fig. 80. Lavoisier and his young wife in the laboratory. In the centre is seen his famous apparatus for demonstrating the composition of air.

After 1774 both hydrogen and oxygen were known and several chemists exploded mixtures of them with a very fine bang and noticed a deposit of moisture on the vessel. Cavendish in particular carried out this experiment in the vessel shown in Fig. 79, and by repeatedly refilling it with the gases and exploding it, accumulated an appreciable quantity of water. Why could he not simply say 'two parts of hydrogen and one of oxygen combine to form water'? The first difficulty was that he and many others were convinced that water was an element and was already present in the oxygen or hydrogen, which they did not suppose to be elements.

While these discussions were in progress Lavoisier heard of Cavendish's experiments of producing water from 'oxygen' and 'hydrogen' and saw the meaning of them. Oxygen and hydrogen were elements and water was the compound of them.

So chemistry was purged of the relics of Aristotle and the alchemists, and made a new beginning on the objective basis of what could be observed, isolated, weighed and measured. Only a few years had to pass before Dalton's atomic theory made it the science it is today.

CHAPTER FOUR

THE REFOUNDING OF THE SCIENCES

THE FIRST QUARTER OF THE nineteenth century, the time of the Napoleonic Wars and ten years or so after, was a great age of science; and if we are to sum it up in a phrase we may call it the time in which the scientists began to understand the nature of what they were investigating. It was the period when they brought convincing evidence that matter was made up of atoms and molecules and that light and other newly discovered radiations were waves in an all-pervading ether. Of course, many other first-rate discoveries were made that did not bear on these fundamentals.

First, as to atoms and molecules. Scientists had believed for a long time that matter was made up of particles or atoms but they did not make use of the idea to explain their results and predict others. It was John Dalton, an English Quaker, who by making some new assumptions about atoms explained the facts of chemical combination, and so convinced the world that the theory of atoms was true and important. His assumptions were simple. Every chemical element (as defined by Boyle and Lavoisier) was entirely made up of its own particular kind of atom different from any other, and all the atoms of any one element were identical, especially in weight. Chemical compounds were entirely made up of identical particles which we call *molecules*, each composed of a fixed and invariable number of atoms of different kinds. If this were true it meant that a particular compound always contained the same proportion of the component elements; that if there existed two compounds of the same elements, the quantities of one of these elements that combined with a given quantity of the other would be present in the proportion of whole numbers, as 1:2, 2:3, etc. Dalton showed by experiment that this was true, and so concluded that his theories about atoms and molecules were correct.

Thus he originated the central notion of modern chemistry—that the important thing to know about a chemical compound was the number and kinds of atoms in its molecule, and he introduced the *chemical formula* to give a picture of it. On p. 106 is one of Dalton's formulae, and beside it are two modern representations of the same chemical molecule. Dalton's formulae were intended to show the number and kind of atoms in the molecule but not the way in which they were arranged; for this science had to wait some fifty years. Dalton's central idea was right, but a tremendous amount of work was needed before his chemical formulae could be proved or disproved; and, although he put out his ideas in 1808, the chemists were not agreed on their formulae till

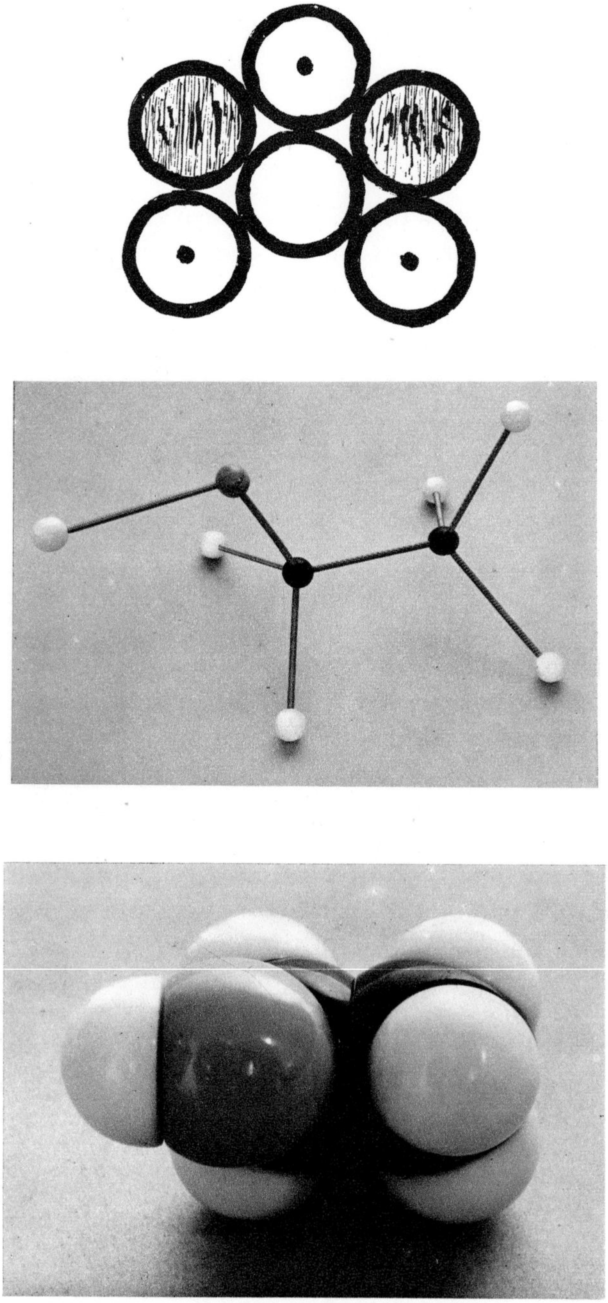

Fig. 81. *Above,* Dalton's formula for alcohol: *middle,* a modern model showing the disposition of the atoms in space: *below,* a model indicating not only this but their relative sizes.

Fig. 82. 'Thou knows it must be so, for no man can split an atom.' The Quaker, John Dalton, gives instruction to two of his pupils, one of whom may be imagined to be the youthful J. P. Joule.

about 1860. But, for all that, the chemist's aim, from the time of Dalton to the present day, has been to find out the formulae of compounds and to discover how their properties and behaviour depend on their formulae.

So Dalton and the other chemists who followed him were soon quite convinced of the existence of atoms and molecules: at the same time the physicists were beginning to find them equally important in their attempt to answer the question, 'What is heat?'

Now this was a very important question, because, throughout the eighteenth century, more and more steam-engines had been turning heat into valuable work, used for pumping water and more recently for driving machinery. To run these engines men had to burn coal; coal and its transport cost money: so engineers wanted to understand heat so as to be able to use it to the best advantage.

There had for a long time been two theories of heat. The usual one was that it was a sort of matter—tiny material particles that could penetrate through and into bodies: it was in fact listed as a chemical element by Lavoisier and named 'caloric'. As a hot body weighed no more than a cold one, caloric had to be treated as weightless, a property in which it differed notably from any other element. The prime difficulty about this theory was to explain the production of heat by friction. Those who believed in 'caloric' thought that it was contained in everything in a sort of latent or dormant state, and that, when two things were rubbed together, caloric was squeezed out, like water from a sponge, whereupon it became apparent as sensible heat. Benjamin Thompson (Count Rumford), who founded the Royal Institution, and Sir Humphry Davy, its first director, held the opposite view. Rumford is remembered for a very striking experimental test of the two views. He had observed that cannon became hot when they were being bored. So he attached to the cannon a cylinder of metal and enclosed this in a wooden box. When the metal was rotated by the power of a horse-mill against the friction of a blunt borer, enough heat was evolved to boil several gallons of water (Fig. 84). The metal of the cannon and the borings detached from it were quite unaltered; so the heat could not have been extracted from these. Unlimited heat could be got out of the apparatus, nothing but motion was put into it: it followed that heat was a sort of motion, presumably of the atoms or molecules of the hot body. This theory made the study of heat into a problem in dynamics—already a well-developed science. Heat was the energy of motion of particles: the next stage was a theory as to how the particles moved, but this belongs to a rather later period.

The next basic problem to be partly solved was that of the nature of light, whether it was a stream of corpuscles, as it were, minute bullets or a train of waves, like those of sound. Newton thought it must consist of particles, because

Fig. 83. An atmospheric engine of the close of the eighteenth century employed for pumping and winding at a colliery.

Fig. 84. Benjamin Thompson (Count Rumford) demonstrates that an unlimited quantity of heat can be produced by the expenditure of nothing else but motion.

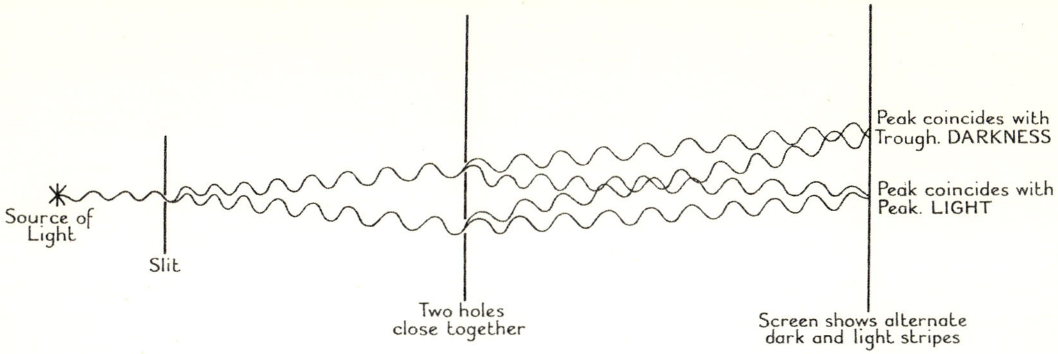

Fig. 85. Illustrating the manner in which two rays of light may meet and cancel each other, so producing darkness.

light did not seem to travel round corners as water-waves or sound-waves do. None the less the phenomenon of Newton's rings was hard to explain on this theory. We have seen that Newton came to a conclusion rather like the modern one, that light had properties both of corpuscles and waves, but during the eighteenth century the corpuscle theory was generally adopted. Soon after 1800 Thomas Young brought strong evidence against it. He made two minute holes in a screen, very close together, and let a beam of sunlight pass through them and fall upon white paper. The rays from each hole spread out into a patch, and where these met the resulting area of light was crossed by a number of dark bands.

If two streams of corpuscles strike the same point on the paper, it should look brighter than if only one did so, whereas two streams of waves can interfere with each other, the peaks of one falling into the troughs of the other and destroying each other. So if two beams of light could make darkness, light should be thought of as waves.

Fig. 85 shows what happened. The rays from each hole travel different distances to meet the screen and sometimes rays, from the two holes, striking the screen at the same point will differ in length by $\frac{1}{2}$, $1\frac{1}{2}$, $2\frac{1}{2}$, etc., wave-lengths: their peaks and troughs will then coincide giving darkness.

This experiment proved convincingly that light consisted of waves, but Young still supposed that these waves moved lengthways (in-and-out) like sound-waves, and did not think of them as moving cross-ways (up-and-down) like water-waves. His work was taken up by many others, notably by the Frenchman, Fresnel, who worked out the mathematics of the wave-theory very completely.

But one odd phenomenon remained unexplained and in the end proved to be the key to the problem.

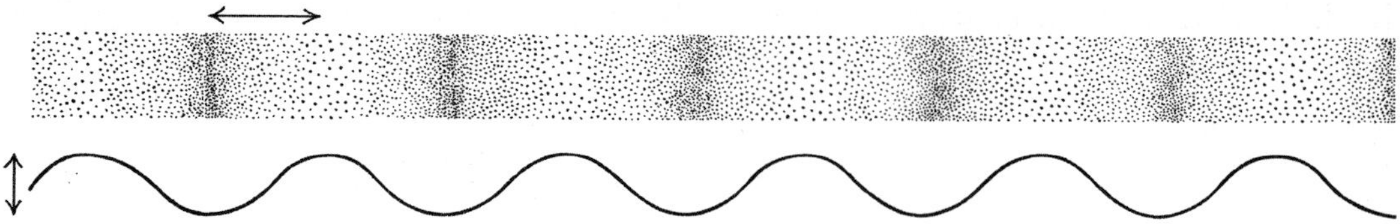

Fig. 86. Diagram of (*above*) longitudinal and (*below*) transverse waves.

Before 1678 Christiaan Huygens had discovered that light could be 'polarised'. If a ray of light passed through certain crystals, of which Iceland spar is the best known, it was found to split into two rays. If one of these rays was then passed through another such crystal, it traversed it undimmed if the crystal were in one position, but was extinguished if the crystal were at right angles to that position. Newton saw that in this case light behaved as if it had sides; as if the ray were a flat lath rather than a round rod. The significance of these experiments was not grasped at that time because no one had thought that light could consist of transverse waves; all through the eighteenth century the corpuscular theory held the field, and so no further attention was given to the matter. But in 1809 Malus discovered that light could be polarised by mere reflection and Sir David Brewster and also Fresnel worked out the laws of polarised light very thoroughly. But the very existence of polarisation seemed to be incompatible with the wave-theory as long as it was based on longitudinal waves, and it was Thomas Young, once more, who led the way by suggesting that the waves of light were transverse. This explained the fact that a beam of polarised light could be turned through an angle (as a key is turned in a lock). This turning can make no difference to the longitudinal waves, but the plane of the peaks and troughs of the transverse waves, when twisted, will present a different angle to anything they pass through. The point can be illustrated by a very simple experiment.

One of the easiest ways to polarise light is to pass it through a slice of the transparent mineral tourmaline. This, we now know, selects from the light only those waves that make a particular angle with the pattern of the crystal. We may think of the tourmaline crystal as a sort of grid-iron that lets through only the waves that are parallel to its bars. Now if the polarized beam strikes another tourmaline it can only pass if the bars of its 'grid-iron' are parallel to those of the first. Now we can put between the two tourmalines a cell containing transparent sugar solution, and the light is suddenly cut off by the second tourmaline. The sugar has twisting molecules and these twist the light so that it can no longer pass the second 'grid-iron'.

Experiments like these proved that the twisting of light made a difference in its behaviour. It was thus proved to be made up of transverse waves like water-waves, and this idea soon solved a great many problems about it, though it still left open the question of what was vibrating and under what kind of force, a problem not to be solved for another half-century.

Perhaps the most exciting discoveries of the early nineteenth century were those that followed from the invention of the electric battery.

All the electrical experiments of the eighteenth century were done with the very minute quantities of electricity that frictional machines will provide, and the effects, magnetic, thermal and chemical, which depend on a large current of electricity, were scarcely noticed. Thus when Volta constructed his battery, which gave a plentiful supply of low-tension electricity, a wide avenue of scientific progress was opened. The discovery of the electric battery was an Italian one. It begins with Luigi Galvani's discovery in 1786 that frog's legs, isolated from the rest of the animal, could be made to contract by electricity from a frictional machine or Leyden jar. On one occasion the legs were hung from a copper hook attached to an iron railing: when the legs touched the railing they contracted without the application of electricity (Fig. 87). Galvani saw that electricity was being produced by the frog's legs and the two different metals. He did not, however, fully understand that the electricity was produced by the metals and not by the frog, but Alessandro Volta, who was much more of an electrician and had already discovered the electrophorus, followed up his work and in 1799 discovered the Voltaic battery, the essence of which was *two different metals in contact with* a *conducting solution.* In Fig. 88 can be seen two of the earliest forms of the battery. Volta himself did not do much with it, but it rapidly developed into a most powerful instrument of research.

The first remarkable effect of the electric current to be discovered was electrolysis, the breaking up of chemical compounds into their elements, and the first important example of it was the electrolysis of water by Nicholson and Carlisle in 1801. Their battery was a pile made up of a hundred silver discs (half-crowns in fact), a hundred zinc discs of the same size and a hundred pieces of green baize soaked in salt solution. These were piled up in the order zinc, silver, baize, zinc, silver, baize, zinc, silver. The two end discs were connected through water (for some reason) to two platinum wires in small tubes containing dilute acid; bubbles rose from the wires and hydrogen and oxygen were collected. This gave the hint that the electric current might decompose some chemicals that could not yet be decomposed into their elements, and led to Sir Humphry Davy's great discovery of the metals sodium and potassium, which he made by electrolysing caustic soda and caustic potash. The chemists had recognized that the great class of salts, including such materials as silver nitrate, copper sulphate, common salt, saltpetre, were compounds of acids and bases and that bases were compounds of acids with the oxides of metals.

Fig. 87. Luigi Galvani hangs a preparation of a frog's legs by a copper hook to an iron railing and is surprised to see them contract in the same way as they would under the shock of an electric machine or Leyden jar.

But there were still many salts from which no metal had been extracted and the most important of these were the very well known and useful salts formed by the action of acids upon soda and potash. The action of the electric current upon solutions of these salts did not produce the metal, but only the base, so Davy decided to try the effect of a very powerful electric current upon the solid bases. The Voltaic battery gave a good current when first connected, but this rapidly diminished to a very low figure. There were, however, many other combinations of metals and solutions that would yield electricity and other types of battery were soon invented. Davy, at the Royal Institution, constructed several enormous batteries with thousands of cells, which, alas, have long since disappeared. His discovery of sodium and potassium was made with such a battery. This fact makes it difficult to repeat his experiments, for we cannot get this type of current from our mains. He put a piece of caustic soda in a platinum cup connected to one pole and touched it with a platinum wire connected to the other: it melted and globules of melted sodium metal were formed and swam to the top and burned. So the alkalies were proved to be compounds and the world learnt of these two extraordinary metals, soft as putty, light enough to float on water and catching on fire when thrown upon it. In spite of this, electrolysis did not come into use for electroplating until 1840, the key invention being the use of cyanides in the plating bath.

The next discovery to astonish the world was the electric light, the arc-light, also discovered by Davy. Again he used his very large battery and connected to it two bits of charcoal (Fig. 89). He touched them and separated them, whereupon a brilliant arch of such light as the world had never seen bridged the gap between them. This discovery also lay dormant for a long while simply because there was no way to obtain a continuous supply of electricity at a reasonable cost until 'electromagnetic machines', that is to say, dynamos or generators, had been perfected.

The most influential electrical discovery of the period was that of the connection between electricity and magnetism. There had been hints of this: thus lightning and powerful discharges from Leyden jars had been known to magnetize iron, but it was not until twenty years after the electric battery was invented that the current was found to have magnetic effects. Earlier experimenters had not realized that the current would need to be flowing and expected an effect from a wire merely electrified by being connected to a pole of the battery: this was, of course, a hang-over from the technique used with the high-tension machines of the eighteenth century. Moreover the needle would be expected to move to or from the wire, not across it. In 1820 the Danish physicist John Christian Oersted discovered that when an electric current was passed through a wire a compass-needle lying parallel to it was deflected

Fig. 88. Alessandro Volta experiments on animal electricity with his new battery and pile.

Fig. 89. Sir Humphry Davy demonstrates the electric arc at the Royal Institution, 1808.

(Fig. 91). It is said that the discovery was made by accident during a lecture, but in fact he had been on the look-out for such an effect for many years. At once he set to work to investigate the phenomena and was able to show that the influence of the wire would pass through all manner of substances interposed between it and the needle. The discovery did not have to wait for recognition for within a few days Ampère, in France, stated the laws governing this effect and Arago, in the same country, showed that a wire carrying a current would attract iron filings. Thus it was established that an electric current was associated with a magnetic field. Five years later William Sturgeon, a Lancashire man, made the first electromagnet, the invention at the root of all the telegraphs, telephones, generators, motors and the like that have transformed our way of living. Joseph Henry in the U.S.A. thought of insulating wire with silk and using many turns of it: in this way he made magnets able to support a ton.

Fig. 90. Faraday's large electromagnet preserved at the Royal Institution.

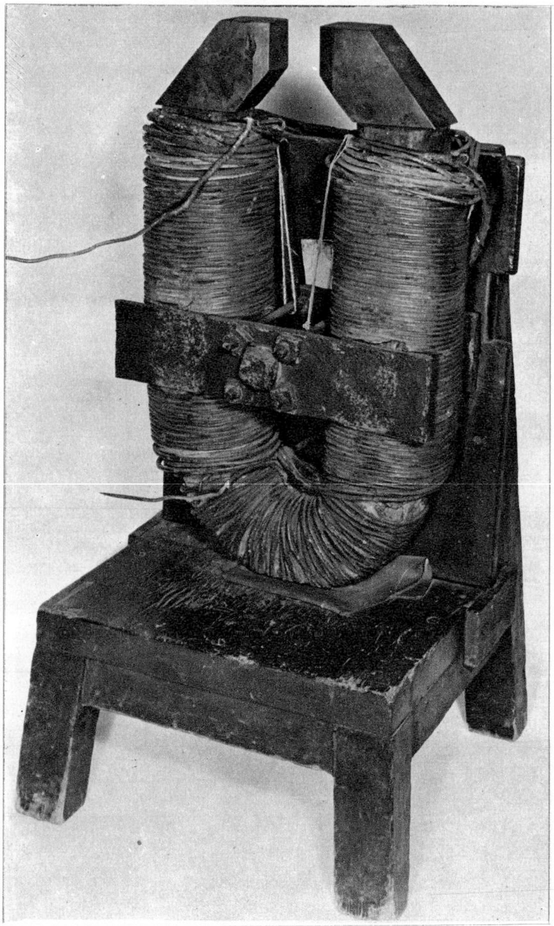

The electromagnets so made were far more powerful than loadstones or the magnets made by stroking steel with loadstone, and so they gave the opportunity of finding out more about magnetism. Fig. 90 shows Faraday's electromagnet, which is preserved at the Royal Institution, where it was used. With this magnet Faraday was able to make two fundamental discoveries. The first was that almost all elements were attracted or repelled by the magnet, though in far less degree than iron: this showed magnetism to be a general property of matter, not merely a special property of iron and one or two kindred elements. The second was the discovery of the effect of magnetism on light, which helped to prove light to be an electromagnetic wave. This epoch-making experiment is described in the next chapter of this book.

Fig. 91. H. C. Oersted succeeds in deflecting the compass needle by means of a strong electric current.

Now let us turn to something quite new. Up to this time science had been studying what the things of the present were and how they behaved, but gradually there came into view a new possibility—that of the discovery of the remote past. Men had so far been content with history—with what was written in books, which did not claim to go back more than six thousand years, but when they began to study the earth they lived on, signs were found of an antiquity that made the whole of human history seem very recent.

First of all, men began to go beyond the account of the creation of the world, set down in the book of Genesis, and ask how a body like the earth might have been formed by the operation of those laws of nature that had been discovered by science. Two very great philosophers, René Descartes (in the sixteen-forties) and Gottfried Wilhelm Leibniz about 1700, thought the earth might once have been, like the sun, a white-hot body and had since cooled down, solidified and become the abode of life. Fossils, the remains of living beings—shells, bones and the like, embedded in the rocks—had puzzled men since the time of the Greeks. How was it, for example, that obvious oyster-shells were found embedded in the rocks of the hill-tops, far inland? In the sixteenth century several authors gave pictures of these 'formed stones': some thought they were the remains of living creatures: others that they had been formed in the rock by some such power as that which had shaped the crystals of minerals. Robert Hooke, before 1700, published a fine picture of ammonites,

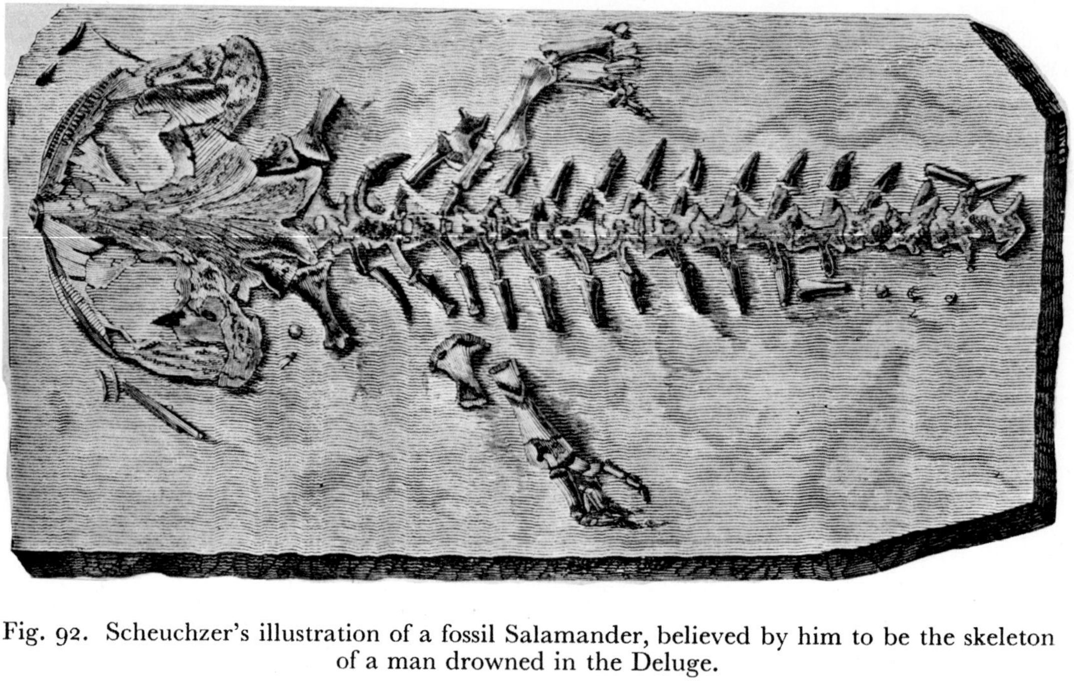

Fig. 92. Scheuchzer's illustration of a fossil Salamander, believed by him to be the skeleton of a man drowned in the Deluge.

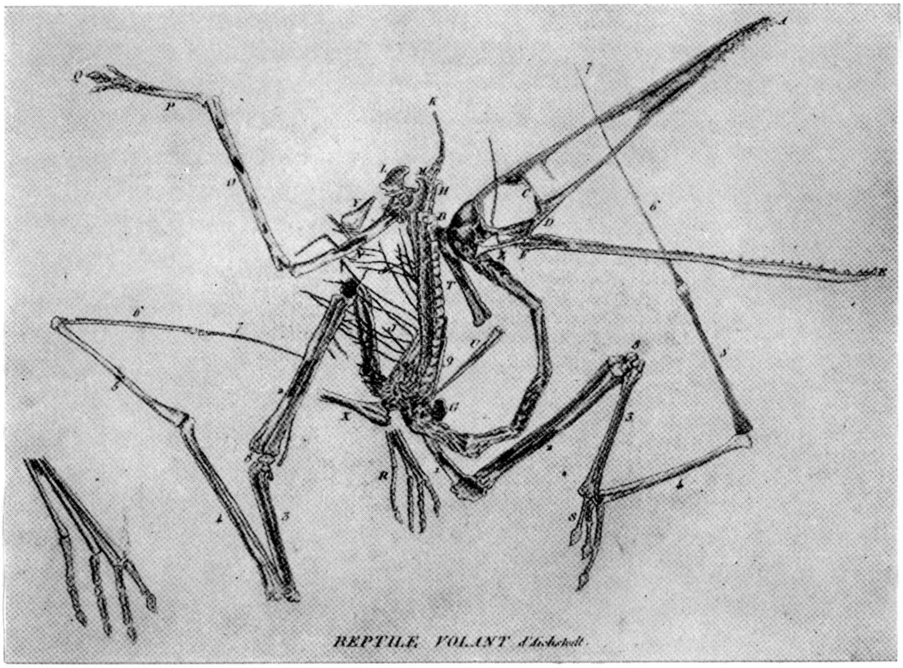

Fig. 93. Baron Cuvier's illustration of the fossil which he recognized as that of a flying lizard, to which he gave the name *Ptérodactyle*.

recognized them as remains of extinct animals, and supposed they had been buried by earthquakes—a hypothesis not far from the truth. But the study of the anatomy of animals other than man had hardly begun, and so it was not borne in on geologists that these fossil remains were of creatures that no longer lived on earth. Some ludicrous mistakes were made: thus in 1731 Scheuchzer took a fossil of a kind of giant Salamander (Fig. 92) to be the remains of a man and regarded it as evidence for the reality of the Biblical deluge.

But during the eighteenth century the study of the anatomy of animals began to throw light on these fossils. It gradually became clear that the earth had formerly supported a population of animals quite unlike its present inhabitants. It also became evident that the earth was very old. The six thousand years of recorded history gave no hint of these ancient organisms, and men began to talk of sixty or a hundred thousand years ago as a probable date for the beginnings of things. Later, Baron Cuvier, in the years between 1800 and 1830, made knowledge of these past forms of life much more precise. He became so expert a comparative anatomist as often to be able to inspect a single bone and name the kind of animal it came from. Fig. 93 is his illustration of the fossil bones of a pterodactyl: previous scientists had attributed these to a bird or a mammal, but Cuvier saw they were those of a flying lizard.

Such works as Cuvier's introduced the burning question of how these animals became extinct and how new forms took their place. The first answer was that they were drowned in the Biblical flood, but it was soon clear that one flood was not enough, and so a series of violent catastrophes were supposed, each blotting out a population, the earth being re-peopled from survivors. This view was the usual one, but another school of thought, advanced by the Scotsman James Hutton, about 1795, thought it unscientific to assume these vast deluges and cataclysms of a type which did not appear to have occurred in historic times; he thought the only legitimate way to account for the mountains, valleys, seas and the like was to suppose they had been produced by the processes we see going on today, namely, erosion by rivers and seas, and the slow rise or fall of the surface of the earth. If this were accepted, the earth must have taken so vast a time to come to its present state that Hutton simply said he could see 'no vestige of a beginning, no prospect of an end'.

Eighteenth-century geology had all the eccentricity of a science with too much theorizing and too few facts, but in the early years of the nineteenth century the geologists made serious attempts to make maps that showed the positions of the layers of different kinds of rock. These studies made it necessary to suppose that a gigantic depth of rock had been laid down from water and that an enormous time must have elapsed while this was going on.

So in this first quarter of the nineteenth century men came to realize that the earth was very ancient and that in the remote past it was peopled by very strange beings, but how these had died out and the present animal population had come to be, remained obscure. One or two men, for instance Erasmus Darwin, the grandfather of Charles Darwin, and the Frenchman Lamarck had suggested that the living beings of the past had been transformed into those of the present by a long series of slow changes. This theory became pretty well known, but was not taken seriously until the eighteen-fifties.

If the earth were old, what of the solar system and the stars? Could science conjecture how the universe had come to be? Again a great philosopher, Immanuel Kant, started the quest by suggesting that the whole universe might have been formed from a uniform cloud of cosmic dust. The first serious scientific attempt to explain the existence of the solar system was that of the French astronomer and mathematician P. S. de la Place. He supposed that the sun and planets had been formed from a vast cloud of glowing nebulous matter in slow rotation. As the mass cooled it contracted; but as the total amount of rotation remained the same, the cloud had to spin faster and faster, with the result that the velocity of its surface became so great as to balance the contrary force of gravity. A ring of glowing matter was thus separated from the equator of the mass and this ring—rather like one of the rings of Saturn—broke up into fragments which finally joined up to form a planet. This process was then repeated so as to form all the successive planets, while the residue formed the sun at the centre. This theory was good enough to last a century: we no longer believe in it, because we know that there is not now enough, and presumably never was enough, rotation in the solar system to work it, and because such a ring, if formed, would not condense into a single planet. Modern theories have to bring in a star or even two stars, approaching the sun, to do the work! But la Place's ideas were provisionally accepted and the scientists of the early nineteenth century had to allow a vast period of time, first for the solar system to form, then for the earth to cool, and, finally, for the various past populations of the earth to have arisen, become extinct and been replaced by others. To pass from the notion of long periods to even approximate dates was another problem, the solution of which began after 1850 and, though it has progressed greatly in the subsequent century, is still a living subject of research.

Thus far, little has been said about the essential question of biology—the nature of life—because science had not yet found any experimental approach to the problem. The men of the eighteenth and early nineteenth centuries were first-rate anatomists and had mapped out every organ, muscle and nerve that they could see in man, and had made fruitful comparisons between the anatomies of the various types of animals: about the life-process, however, they knew very little. Microscopes had as yet poor definition at high magnification and the scientists could not see the fine details of the body, and were not aware of the finest details of its structure.

Yet there were advances in the study of the chemistry of living things. Lavoisier had shown that respiration was oxidation, the burning of the substance of the body, and he made the significant statement, 'Life is a chemical function.' No doubt it is much more, but it is certainly that. Thus he proposed as the goal of discovery not only the shapes and sizes of the organs and vessels of the animal, but the chemical changes that went on there. The chemistry of living things was so different from the simple chemistry of metals and acids and salts, that men doubted whether it could be described by the same laws that were being discovered in the laboratory, and many thought that some 'vital force' entered into the making of the fat, sugar, albumen and many other compounds that were found only in living things. Wohler's preparation of urea, an animal product, from ammonium cyanate, an inorganic product, seemed to settle the matter, and its author proclaimed the possibility of making in the laboratory all the compounds found in animals and plants. His proof was really a faulty one, but his conclusion holds good. So in the period from 1780 to 1830 the idea of biochemistry began, though its progress had to wait until chemistry and its techniques had had time to develop.

CHAPTER FIVE

THE BEGINNING OF THE HARVEST

THE SCIENTISTS WHO WORKED BEFORE the eighteen-thirties had spent not a little of their energies in disproving archaic theories and uprooting fixed ideas that had been inherited from the ancients. They had with much difficulty learned the best way to go about their business of research, and their stock of knowledge and technique was steadily accumulating. Nothing in science is lost; the men of each age possess all the equipment of the ages before them. For this reason the progress of science always accelerates, and in the years between 1830 and 1900, science challenged every other avenue of knowledge and established itself as the foundation of industry.

The first major scientific idea that was grasped in this period was that of *energy*. What we explain in terms of energy, was explained in the eighteenth century in terms of material substances: light consisted of material particles; hot bodies contained the element of caloric; magnetized bodies, the magnetic effluvia; the living animal, vital spirits. In the eighteenth century there was no reason to suppose that these had any more relation to each other than had, for example, the chemical elements, but as the nineteenth century progressed it became clear that energy of motion (*vis viva*), heat, light, electricity, chemical action and magnetism could be converted one into another and could be thought of as manifestations of one entity.

We have seen that in the early years of the nineteenth century it was shown that motion could be converted into heat, and also that electricity could give rise to magnetism. A further step was taken when magnetism and motion were made to produce electricity.

Fig. 94. The lines of force of a magnet, outlined by means of iron filings.

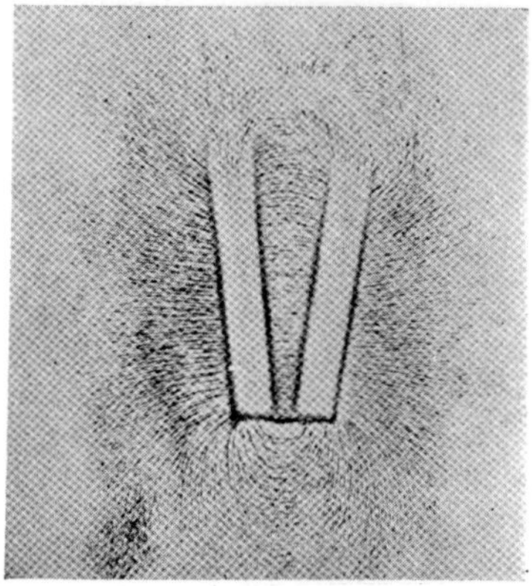

Perhaps the most famous and important of all such experiments was that by which Michael Faraday, in the Royal Institution, produced an electric current by moving a conductor so as to cut the lines of force of a magnetic field. The 'lines of force' represent the paths along which the pole of a magnet impels the pole of another magnet. Faraday illustrated their course very well by means of iron filings (Fig. 94). His great discovery was that if a conducting

Fig. 95. Michael Faraday pulls a bar-magnet out of a coil of wire and so produces an electric current, detected by the astatic galvanometer.

wire moved so as to cut these lines of force, a current flowed through it: the same of course happened when the magnet generating the lines of force was moved and the conductor kept still. Fig. 95 shows him pulling a bar magnet out of a coil of wire, so setting a current of electricity moving through the wire and galvanometer. Faraday's interest was in the relation between electricity and magnetism, but his simple experiment was the main root of the modern electrical industry. Electricity could be made by moving wires in the field of magnets. Thus instead of using up expensive metals, such as zinc, in order to make electricity, it was enough to use up coal in a steam-engine and use it to turn coils of wire in the fields of magnets. Almost at once electrical generators began to be constructed and slowly increased in efficiency: their first use was to provide the large currents required for electroplating. A decisive advance was the replacement of the permanent magnets by electromagnets energized by the electricity that the generator produced. This was the essence of the Siemens dynamo, yet there were many difficulties in making the dynamo efficient enough to supply electricity at a cost low enough for use for lighting and power: the work of T. A. Edison about 1878 was the final step, and from that year the electrical age must date. The distribution of electrical energy from central stations began, and with increasing demand for electric lighting and, later, electric traction, the cost of it steadily diminished. New uses were found, new difficulties were solved by research, until today there is scarcely any industry or branch of science that is not dependent upon a supply of electricity.

Let us return to the idea of energy. By the eighteen-forties scientists and technologists had found that heat, electricity, magnetism, chemical change and energy of motion, could each be converted into the other, and that according to a fixed tariff, so to speak. At first men talked about the 'equivalence of the powers of nature'; then they boldly asserted that all these were manifestations of the same thing, the capacity for doing work, to which, following Thomas Young, they gave the name of *energy*. That was a wonderful simplification of the world, for it now seemed that the scientist was talking about only two things, matter and energy.

An English scientist, J. P. Joule, worked out just how much heat could be obtained from a given amount of work. He constructed an apparatus in which weights (like those of a grandfather clock) turned a paddlewheel immersed in water, which became very slightly warmer as the result of its friction. The weights and the distance they fell measured the work done, the rise in temperature of the water measured the heat produced. Very many precautions had to be taken, but Joule derived from it the fact that 772 foot-pounds of work were needed to produce enough heat to raise the temperature of a pound of water by 1 degree F. He also tried a much simpler method, to measure the temperature

Fig. 96. The first highly efficient type of dynamo, designed by T. A. Edison.

at the top of a waterfall and at the bottom: the work done by the falling water made it slightly warmer, and the quantities of work done and heat produced are easily calculated from the height of the fall and the temperature-difference.

The contrary problem—how much work you can get out of a given quantity of heat—was not so simple, because the engines which turn heat into work can never convert into work more than a certain proportion of the heat in the working substance (e.g. the steam in the steam-engine). But the problem of calculating the highest possible efficiency of various types of engines was worked out about 1824 and became vastly important to engineers.

All this work led up to one of the greatest of all scientific generalizations, the law of conservation of energy, first expressed by a physiologist, Robert Mayer, but most clearly put by von Helmholtz in 1847. This law states that energy is not created or destroyed but merely changed from one form to another. Lavoisier had already announced the conservation of matter, assumed in all chemical work, and for sixty years the world seemed to be a closed system, with an unalterable quantity of matter and an unalterable quantity of energy. From 1906, however, advanced spirits began to recognize the possibility of turning matter into energy, while today the continuous creation of matter is a much-discussed hypothesis.

One interesting consequence of the idea of conservation of energy was a new approach to the problem of the age of the universe. If the sun were no more than a hot body radiating its heat-energy, cooling like a red-hot poker, it could have maintained its heat only for a few centuries; but in fact the geological record gave no reason to suppose that the power of its rays had diminished in the last 100,000 years and perhaps far longer. Von Helmholtz overcame this difficulty by showing that as the sun slowly contracted under the pull of its own gravitation, the work done by its vast mass falling towards its centre was equivalent to enough heat to have maintained the present rate of supply for twenty million years, which seemed at that time to be long enough for the changes that the geologists supposed to have happened, though it was not really a hundredth of what we now believe to be the earth's age, let alone the sun's.

And this may take us on to the great problem of the nature of that which we now call radiation. Light had been studied for thousands of years, but no one had proved that there existed what we may call 'invisible light' until the years around 1800–1. Only then was the spectrum of the sun, the band of colours that Newton's prism had produced, studied by other means than the human eye. Sir William Herschel, the great astronomer, discoverer of the planet Uranus, placed thermometers at various points in and beyond the visible spectrum and found that beyond its red end where nothing was to be seen, there were rays that heated a thermometer bulb even more than did the visible coloured light.

Fig. 97. William Thomson, Lord Kelvin, walking in Switzerland, finds J. P. Joule measuring the temperatures of waterfalls with a very long and sensitive thermometer, while his bride waits in the carriage.

About the same time the first studies of photography were being made: these were concerned with the darkening of silver salts by light and it was found that rays beyond the other end of the visible spectrum—the dark part beyond the violet—darkened the silver salts more rapidly than did the bright rays. It was clear that the sun's radiation contained kinds of 'light' other than those that man can see.

Next came the investigation of what light was. As we have already seen, soon after 1800 it was proved to be a wave-motion and therefore a kind of energy, not a kind of matter. But the idea of a wave seemed to require a medium that could undulate or *waggle*. For at least two thousand years men had believed in an ether, a sort of subtle semi-material medium, that penetrated all things and filled the whole of space. This ancient cult-object had been made to do duty as a means of transmitting forces; now it was to be the medium in which light-waves moved. Since light was a transverse wave, the ether had to be thought of as able to transmit this kind of wave, and to be pictured as something that could *waggle sideways*. Now a gas cannot move in this way but a solid can, so that the men of the nineteenth century had to think of the ether as a sort of elastic solid or jelly. But the stiffer the jelly the faster the waggle, and since light vibrates billions of times a second the ether had to be thought of as a very stiff jelly indeed—much stronger in fact than steel. Yet the atoms and molecules were all the time moving through it quite undisturbed! That the ether had to be thought of as a very odd substance did not disprove its existence, but it led men to look for a simpler view.

A new way of thinking about light arose from the proof that it had something to do with magnetism and electricity. The experimental proof, once again, was the work of that great man, Michael Faraday, who showed that a magnetic field could cause a ray of polarized light to rotate: if a magnet could affect light, there ought to be something magnetic about it. Then came the remarkable fact that the velocity of light could be calculated from purely electrical and magnetic experiments in which light was not used at all! From evidence of this kind the physicist, James Clerk Maxwell, propounded the theory that light was a rapid transverse alternation of electrical and magnetic

Fig. 98. Faraday's sketch of the way in which he twisted a beam of polarized light by means of a magnetic field.

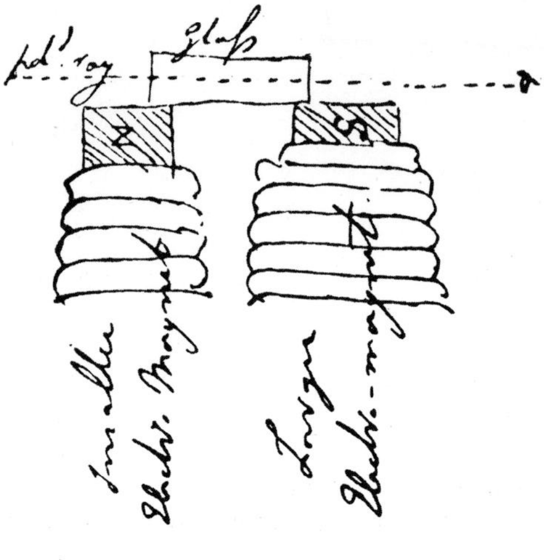

Fig. 99. Sir William Herschel measures the effect of different regions in and beyond the visible spectrum on thermometers. Caroline, his devoted sister, ministers to his needs.

fields at right angles to each other. Clerk Maxwell showed that in theory wave-motions of this kind could be set up by making electric charges vibrate and that such waves could be of much greater or much less frequency than the waves of light. In 1887 Hertz set out to prove it. He set up his vibrations

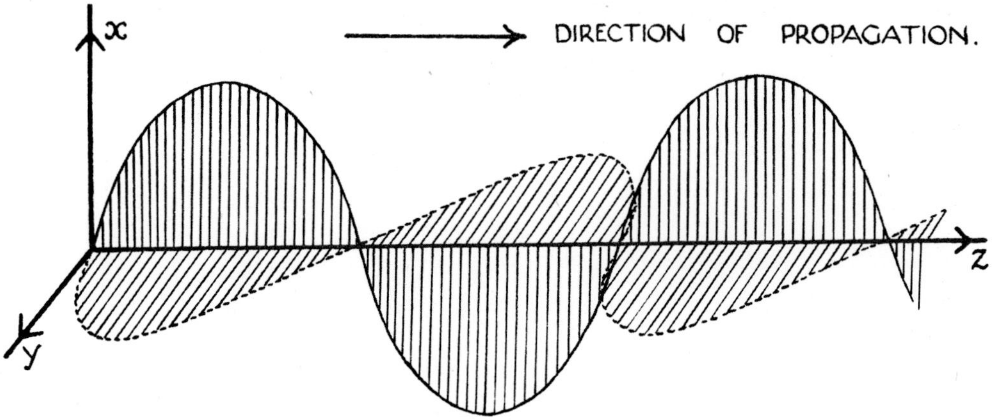

Fig. 100. Diagram illustrating an electromagnetic wave. The wave-lines are *graphs* showing how the strength and direction of the electric and magnetic forces vary: in the actual radiation there is nothing shaped like a wave.

by discharging a condenser across a spark-gap and he showed that waves were produced by receiving them with a rather similar arrangement in which they produced a tiny spark, visible only at very close quarters. He proved they behaved like light-waves by refracting them with a big prism made of pitch. Such was the beginning of radio, though it was some seven years before anyone consciously used these waves to send a message.

Hertzian waves were of great length and low frequency compared to those of light, but waves of much higher frequency than light were soon to be discovered. These, and much else that was important, were discovered through experiments on the remarkable appearances that were manifest when electricity passed through gases at low pressures. Once more serious work on this began with Faraday. He passed an electric discharge through an exhausted bell-jar and showed that the sparks produced at atmospheric pressure were replaced by a steady glow. But the air-pumps of Faraday's time would at best remove all but a thousandth part of the air from a receiver, and the really interesting things happened only when the pressure of gas was reduced still further. About 1860 a much better type of pump, operated by mercury instead of by pistons, was invented and by its aid vessels could be exhausted much more completely. Moreover, the induction coil, discovered by Ruhmkorff in 1850, gave a better source of high-tension electricity than the frictional machines previously used.

Fig. 101. Heinrich Hertz (1888) generates the first radio-waves by passing sparks between brass balls connected by wires to larger brass balls, and detects the waves by their production of a tiny spark between two balls very close together and connected by a wire ring.

In 1879 William Crookes, a man most fertile in ideas, tried the experiment of passing an electric discharge through a tube so highly evacuated that scarcely any air remained. The electricity passed, but in the form of a stream of minute particles (which we know now to be electrons). The rays were called cathode rays and the tubes cathode-ray tubes.

The phenomena demonstrated in these tubes will be considered in the next chapter, as the beginning of our knowledge of the structure of the atom. For the present it is enough to note that in 1895 Röntgen noticed that a cathode-ray tube in action could cause a piece of barium platinocyanide to glow, even when the tube was wholly screened by black paper. He saw that this must be due to some new kind of rays, that could pass through some materials opaque to light. These, which were given the provisional title of X-rays, astonished everyone by their power of penetrating through considerable thicknesses of matter. Their use in surgical diagnosis began almost at once, but they proved very puzzling to the physicists. They could not be reflected or refracted like light and it was not certain that they were electromagnetic waves until 1910–11, when they were found to be diffracted by crystals—a discovery which also transformed our understanding of the make-up of solid bodies. Very soon the gamma-rays of radium were recognized as electromagnetic waves and it was only a matter of time to produce waves of every length between the mile or so of radio to the 10^{-12} centimetres of gamma-rays.

Radiation is almost our only messenger from space; all that men knew of stars was their light, and in 1850 it seemed quite impossible that we should ever know what a star was made of. But in 1859 Bunsen and Kirchhoff invented the spectroscope. Rays of the light to be studied were directed through a very narrow slit, spread out by a prism into a spectrum, and viewed by a system of lenses. If the light came from a glowing gas (e.g. a flame or electric spark) the spectrum was seen to contain numerous bright lines, each an image of the slit formed by light of one particular wave-length. Bunsen and Kirchhoff examined these and concluded that each chemical element present in the glowing gas gave rise to its own group of lines, recognizable by their pattern

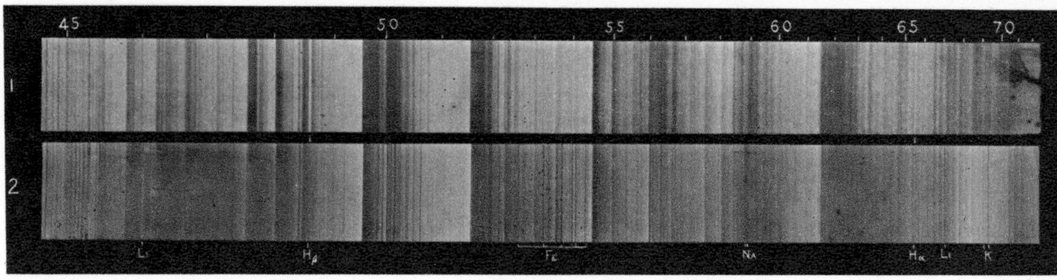

Fig. 102. *Above,* the spectrum of the star Mira Ceti showing the spectrum of titanium oxide and lines due to other elements (Fowler, 1907) : *below,* the spectrum of titanium oxide.

Fig. 103. William Crookes demonstrates that a beam of cathode rays, rendered visible by a fluorescent screen, is deflected by a magnet in the same direction as an electric current would be.

Fig. 104. The earliest photograph of a scientific object, the 40 ft. reflecting telescope at Slough constructed by Sir Frederic William Herschel. This photograph was taken by his son Sir John Frederick William Herschel whose photographic portrait, made by Mrs. J. M. Cameron, in 1865, appears below.

Fig. 105. Robert Bunsen observes the spectrum of a salt through his newly invented spectro scope.

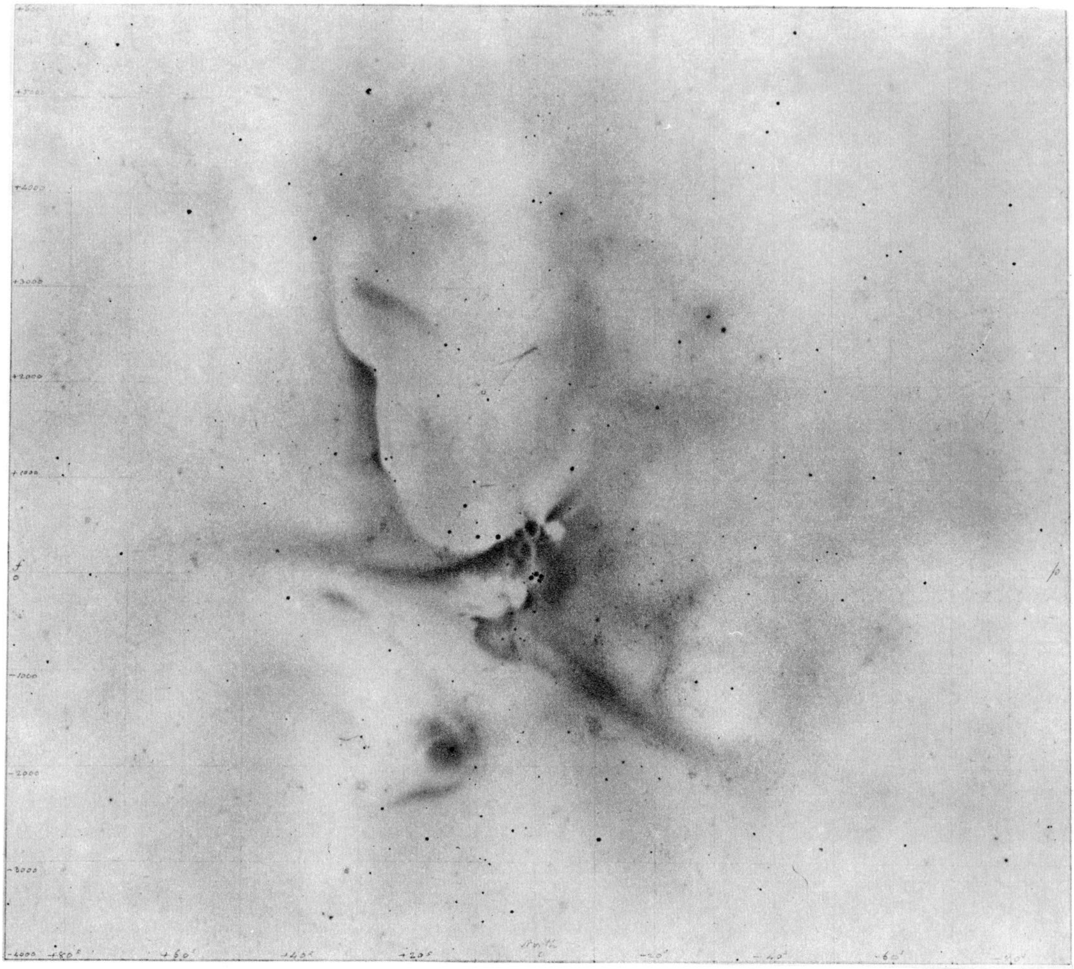

Fig. 106. The nebula in Orion as drawn in pencil by Sir John Herschel.

and position. Light could now tell us what was in the glowing gas from which it came. Bunsen was chiefly interested in it as a tool of chemical analysis, and it soon led to the discovery of several new elements. But Kirchhoff passed light from the sun through his spectroscope: the result (passing over certain complications) was to show the lines characteristic of the elements known on earth. It was not so easy to observe the lines formed in the spectrum of the faint light of the stars, yet when it was done the result was the same and in a few years men had reason to believe that the whole universe was made of the same stuff. Since that time the spectroscope, in its modern forms, has taught us far more: it tells us how hot or heavy the stars are and whether they are moving

Fig. 107. The same nebula as photographed by a modern telescope.

towards or away from us. As a tool of the astronomer it is indeed second only to the telescope.

Another great event for astronomy was the discovery of photography. This science or art developed rather slowly between 1802, when Josiah Wedgwood began to imprint images on leather sensitized by silver nitrate, and 1871, when reasonably fast plates were invented. Thereafter the astronomer no longer had to record each star by viewing and measurement. He could photograph the sky, record a hundred objects at once, and detect any that moved relatively to the others and so left a trace instead of a point: moreover the photographic plate could accumulate faint light for hours and so detect objects that the eye could never have seen. Figs. 106, 107 illustrate the great Nebula in Orion, first as recorded in a pencil sketch by Sir John Herschel about 1835, then as photographed at the Mount Wilson Observatory in 1920.

In the first half of the century it was gradually accepted that heat was motion. Motion of what? Presumably atoms or molecules. What sort of motion? Did the atoms spin, or rush around colliding with each other, or vibrate about centres? It is the habit of scientists to attack the easiest problem first. The eighteenth century had shown that the physical behaviour of gases can be described by a number of very simple laws: thus all gases expand when heated or contract when compressed in nearly the same way: when they combine chemically they do so in simple proportions by volume like 1 : 2 or 2 : 3. So after 1850, when our ideas about energy had been cleared up, a number of scientists decided to assume that a gas consists of perfectly elastic particles (atoms or molecules), moving very fast and quite at random, at an average distance from each other very large compared with their diameter and having no mutual forces of attraction or repulsion. They called this imaginary system a 'perfect gas' and showed that it must do just what gases in general are known to do. No one could see or measure the atoms or molecules of a gas, but because this theory admirably explained what a gas did, they thought a gas must be such an assemblage of atoms or molecules moving in such a way, and all subsequent evidence has confirmed it. But although all the real gases behaved very nearly as the 'perfect gas' should, none of them did so exactly. The reason for these small differences was supposed to be that the atoms of real gases were not points but had a certain size, and also did attract or repel each other a little. These differences between the real gases and the perfect gas provided the first clues for investigating the probable sizes of atoms, and the strength of the forces they exert on each other. But this story belongs to the twentieth century.

An important discovery of the nineteenth century was that, just as steam when cooled or sufficiently compressed formed water, so gases when cooled, compressed, or both, condensed to liquids. Faraday, in the years 1823–45, was again the first to make a systematic attempt to liquefy all known gases. He put his trust in cooling and compression. The simplest apparatus he used was a glass tube having the shape of an inverted V (Fig. 108). In one limb he placed materials that would generate the gas, which built up a considerable pressure: the other he immersed in a freezing mixture. A small minority of the known gases, however, could not be liquefied even by the highest pressures and the greatest cold: these were for a time called 'permanent gases'. The condition for their liquefaction was made clear when Andrews showed that for each gas there was a certain temperature above which no pressure, however high, would liquefy it. So it appeared that the problem of liquefying the permanent gases such as air or hydrogen was not that of obtaining enormous pressures, but rather of making gases very cold indeed.

Scientists tried every expedient, but only minute amounts of the 'permanent gases' could be liquefied before 1895, when several men independently

Fig. 108. Faraday succeeds in liquefying chlorine by heating chlorine hydrate in a sealed tube. Dr. J. A. Paris, his biographer, visited the laboratory while this experiment was in progress, and noticing some oil in the tube rallied Faraday on the carelessness of employing soiled vessels. Faraday opened the tube and the oil vanished: next day he sent Paris a laconic note: 'The *oil* you noticed turned out to be liquid chlorine.'

hit on the idea of regenerative cooling. Compress the gas: cool it as far as you can: make it turn some of its remaining heat into work and so become colder still: then use this cold gas to cool the compressed gas already referred to. When this cooled gas does work it will become colder still and can be used to make the compressed gas still colder. Thus the gas will be progressively made colder and colder until it becomes so cold that it liquefies. The liquefaction of air in this way was accomplished in 1895. The problem of handling liquid air was a difficult one, for in an ordinary room everything is 200 degrees above its boiling point. The essential discovery, here, was the vacuum flask, invented by James Dewar to keep the heat out of liquid air, and today universally employed to keep the heat in tea and coffee.

The first importance of liquid air was that it enabled us to cool all kinds of matter to about −190 degrees C. instead of to only about −50 degrees C. At these low temperatures many new interesting phenomena appeared. It also provided a means of separating gases by liquefying them and distilling them, and charcoal, cooled in liquid air, proved to be the best way of removing the last traces of air from the evacuated vessels used in so many of the experiments that guided us to an understanding of the atom.

What were the chemists doing all this while? First of all, they discovered many new elements and many new ways of making new compounds, so that the books that chronicled them rapidly increased from single volumes to dozens of tomes. Next they learned how to prove the formulae of compounds, that is, to state confidently how many and what kind of atoms there were in a molecule they had never seen nor hoped to see. It took about fifty years to reach this stage, but when it was reached, about 1860, chemistry leapt forward. Once the true formulae of its chemical compounds were known, organic chemistry (the chemistry of carbon compounds) was seen to be a beautifully orderly and regular science. The organic chemist began to be able to write down a chemical formula and say with a fair expectation of success, 'I will make the compound that has that formula.' From this time came the idea of synthesis, putting together simpler compounds to make more complicated ones, and with it came the industries of synthetic dyes and drugs. Once a good dye was discovered, the chemist could figure out a dozen possible molecules similar to but not identical with the dye's molecule, make compounds having such molecules and get a dozen new dyes! So with drugs. The chemist finds a drug that makes you sleep: he then makes a dozen—or a thousand—allied compounds: it is almost certain that some will have the same effect, and it is very probable that one or two will have advantages over the original.

The chemistry of other elements than carbon did not clear up so quickly. These elements soon mounted up to fifty or sixty, and there seemed to be no rules to tell how many elements there could be or to relate one element with

Fig. 109. James Dewar liquefies air at the Royal Institution. In the background are the compressors. Dewar holds one of the double-walled vacuum flasks, which he invented for the purpose of preserving liquid air from the heat of the surrounding atmosphere.

another. So it was a great advance when the Russian, Mendeleyev, showed that if the elements were arranged in order of the weights of their atoms they fell into 'family groups', the members of which resembled each other. These groups introduced a new order into inorganic chemistry, but until the twentieth century, no one could understand why such groups existed.

The nineteenth century saw the beginnings of an understanding of the fundamental conditions of life; its tool was the microscope and the signal for the advance of biology was the perfection of that instrument, which had been but little improved in the eighteenth century. The wave-theory of light, which became familiar in the first quarter of the nineteenth century, was of much value to optical workers, who calculated the combinations of lenses best fitted to make up microscope objectives that would give sharp pictures at high magnifications. Actually a magnification of ×1,000 shows all that you can see with light: more magnification makes the image larger but proportionately less distinct. The single-lens objectives of 1800 gave images that were curved, reasonably sharp only in the centre and with every detail made hazy by colour-fringes. The lens-designers of the next eighty years gradually cleared out the colour, flattened the field and sharpened up the edges: new ways of lighting were invented to show up detail and the biologists discovered ways of staining transparent structures with dyes so as to make them visible.

The first product of this advance was the notion of the cell as the unit of life. In 1838 and 1839 Schwann and Schleiden respectively announced their theories which amounted to the assertion that every part of every animal and plant was built up of cells. The microscopes of the thirties would not show much of their structure but the cell was gradually acknowledged to be the unit of life, the place where the life-process went on. In the sixties and seventies people talked of it as a mass of formless jelly, or *protoplasm*, and indeed were rather superior about it: since then it has steadily been proved to be more and more complex till we now wonder if a lifetime would suffice to describe a single cell at a single moment. A great discovery of the mid-nineteenth century was that expressed by the German pathologist Virchow in the phrase, 'Every cell comes from a cell.' Growth was seen and understood to involve the division of cells to make new cells.

No less important for science than for human health was the discovery that fermentation, putrefaction and infectious disease were caused by minute living organisms. This wonderful story started very prosaically about 1857 with young Louis Pasteur trying to find out why beer, when being brewed, sometimes became undrinkable or 'unsound'. This led him to discover and prove that living yeast was essential to the fermentation of wort into beer. He went further, and proved that living organisms were the cause of things going bad,

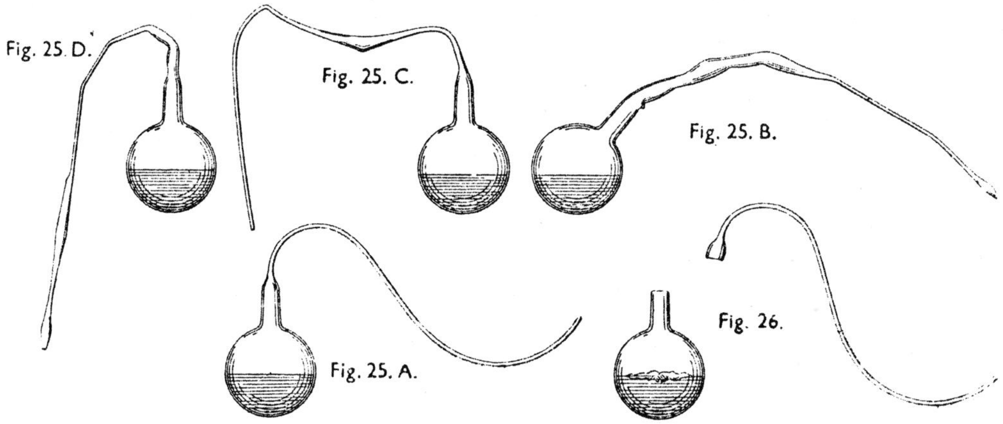

Fig. 110. Some of the flasks which Louis Pasteur used to demonstrate the truth of the germ-theory of putrefaction.

and that in their absence even liquids so easily altered as milk and broth remained permanently fresh. Fig. 110 shows some of the beautifully simple pieces of apparatus that he used. In such a flask Pasteur placed broth, boiled it and left it to cool, and it did not go bad, though weeks and months went by. Air could enter freely but the germs were deposited in the long neck and did not reach the liquid. It was thus proved that air alone could not bring about putrefaction. He then cut off the neck of the flask and the liquid went bad within a few days, for there was nothing to trap the germ-laden particles from the air.

This work of Pasteur also cleared up the great question of spontaneous generation—whether living beings were ever formed from non-living matter. In the seventeenth century there were those who were prepared to say that even such complex organisms as mice could be generated from dirty linen or decaying vegetation. In that century, however, the Italian, Redi, showed that meat allowed to go bad under a wire-gauze cover did not spontaneously breed maggots. But when the microscope showed that all putrefaction was accompanied by the appearance of myriads of minute living organisms, it was very generally thought that these simplest beings were generated from the non-living. Pasteur's long series of experiments killed this belief, and thenceforward, it was taken as certain that all known living organisms were produced from their own kind, and that these minute creatures were the cause and not the effect of putrefaction.

Pasteur's work gave a hint to the Scottish surgeon, Lord Lister, who at that time was trying to discover why the wounds made by surgeons almost always became septic, a process that caused a terrible toll of deaths. He thought the reason was that the fluids in and about the wounds 'went bad' or putrefied. In 1865 he heard of Pasteur's work and followed the clue he had given. If sepsis was putrefaction and putrefaction required living organisms, he would keep them away from the wound or kill them. So he treated his hands, instruments, dressings and the wound itself with carbolic acid, a poison to low forms of life, and he sprayed it into the air. Sepsis, which had killed no less than half of his amputation patients, disappeared. Lister did not bother much about what these organisms were: his job was to kill them. About this time Pasteur's attention was directed to two silkworm diseases that were devastating the French silk industry. Following the hint of his researches on fermentation he found that these diseases were caused by microscopic organisms. Pasteur was not a doctor and would not poach on medical preserves, but it is odd that the physicians who read his work and Lister's did not adopt the hypothesis that what was true for sepsis and silkworm diseases might be true for infectious diseases generally. The belief of the world in 'germs' as causes of diseases dates only from the years about 1878–80, when Pasteur and Robert Koch showed how to cultivate germs and Koch showed how to recognize them under the microscope by staining them with dyes. The discovery that germs are the cause of many diseases has saved millions, perhaps hundreds of millions of lives. It is the reason why we have pure water, proper sewage disposal and generally a hygienic world. Had not these things been found out, a quarter of my readers would already be dead.

Important as these discoveries were, they made less noise in the world than another, the famous theory of evolution, advanced by Charles Darwin and Alfred Russell Wallace in 1858. It put forward the idea that different species of animals may be descended from a common ancestor and have become what they are by a succession of inherited changes. Darwin saw in his hypothesis of natural selection the way to an explanation of the manner in which this might have occurred. The individuals best equipped to preserve themselves from death, and for this or other reasons most likely to be able to mate, would produce the most offspring, which would inherit their qualities; so each generation would become better fitted to survive the struggle for existence. Thus all living organisms were thought to be slowly changing in response to their environment. Darwin held a theory about heredity that has been disproved, but the scientific world still believes that all species have become what they are by the slow changes of evolution. There is still room for argument about the way these changes are guided and brought about, but man's picture of the living world is still based on Darwin's plan.

The greatest shock to public opinion was the suggestion that their own species, *Homo sapiens*, had come into being in the same way and was genetically

Fig. 111. Louis Pasteur lectures on the germ-theory of putrefaction.

descended from a line of creatures now extinct, but presumably more like apes or monkeys than men. The pedigree of man has even today been only very sketchily traced—indeed we cannot point to any fossil and say 'that organism was an ancestor of man'; yet very few people will be found to say that man had not some non-human organism as ancestor. Be this as it may, the theory of evolution became the design upon which the facts of natural history are grouped and it has been the greatest promoter of research in that field for nearly a hundred years. The theory seemed to require an enormously long period for the development of, let us say, man from the single-celled animal. It was impossible to see that any change had taken place in man during the five or ten thousand years of history or archaeology: the twenty million years of past allowed by von Helmholtz's contraction theory seemed all too short for the nineteenth-century evolutionists. Today the dating of the rocks by the 'radium-clock' has given a time scale more than a hundred times as long and possibly enough for the biologists' demands. The building of atoms in the sun and their radioactive decay in the earth's crust affords an explanation of the long and lavish evolution of energy required to keep the earth at a temperature suitable for living beings for a period of some two or three thousand million years.

Fig. 112. The Descent of Man.

CHAPTER SIX

SCIENCE TAKES THE LEAD

IN THE LAST FIFTY YEARS science has taken the lead. By its aid, so much knowledge and power have come into men's heads and hands, that they are ceasing to think of science as some specialized occupation, but are coming to regard it as the standard way of understanding and organizing any job of work. This is indeed a true attitude as regards the understanding and making of material things: yet science still knows little and perhaps will never know much of the human mind: consequently, it can do little for those who see things in relation to men, and men in relation to each other and to God; the artist, the leader of men and the man of religion.

How can the scientific work of the twentieth century be depicted in a few pages, even in the broadest outlines? If I may propose a phrase to express the work of the twentieth century, I will suggest the words 'Down to the Foundations'. By that I mean the successful investigation of the simplest, smallest parts of which everything in the universe seems to be composed, and I also mean the investigation of the simplest things that scientists do and the simplest assumptions that they make.

In the first ten years of the century two very great scientists proposed theories that required a totally new way of looking at the simplest things, and by about 1915–20 everyone came round to their way of thought. First was Max Planck, who proposed the quantum theory, the hypothesis that all energy of vibration existed in minimum portions called quanta, the energy of which was the greater, the higher was the frequency of the vibration. This theory did for energy what the atomic theory had done for matter, and it made sense of many so far inexplicable effects of radiation on matter and *vice-versa*. The second, Albert Einstein, propounded ideas even more revolutionary. He thought about the way in which scientists make their measurements by means of light-signals and the very odd properties they had had to attribute to the ether. In 1905 he advanced his theory of relativity which required the Newtonian laws of dynamics and astronomy to be altered: and in the same year he showed that according to this theory mass and energy could be converted into one another on the basis of a gram of matter to 10^{20} ergs. He thus replaced the law of conservation of matter and the law of conservation of energy by a law of conservation of mass and energy. Furthermore his theory of the world did not require an ether of space and he proposed to do without it: thus getting rid of the last of the weightless fluids which had been so dear to the eighteenth century. He did not just assert these theories but proposed tests for them—and the tests vindicated them and the new advances of science have continued to do so. We must think of Einstein as a refounder of science, perhaps as great as Galileo or Newton. His work is at the basis of the great physical and astronomical discoveries of the century.

Let us now pass from these fundamental theories to achievements more easily visualized. In the twentieth century we learned to weigh, measure and count the atoms of matter; to map out the patterns they assumed; to understand their structure; and, lastly, to alter them and make use of the vast forces that bind their parts together.

The roots of this work go back beyond the twentieth century, to the time when Faraday began to pass electric currents through partially evacuated vessels. This work led on to the discovery of cathode rays by William Crookes in 1879, a discovery which we must now look at more closely. The Crookes or cathode-ray tube was in essence only a highly evacuated glass vessel, furnished with two electrodes, to which a source of high-tension electricity could be applied. Crookes had the services of a highly skilled glassblower, Walter Gimingham, who was able to make him tubes containing various pieces of apparatus which could indicate the properties of the rays. By means of these he proved that the rays moved in straight lines and cast shadows, and that they consisted of heavy particles, for they would turn a little paddle-wheel. They were like an electric current, because they were deflected by a magnet and in the same direction as an electric current would be. They were electrically charged with a negative charge, because they were repelled by a negatively charged plate and attracted by one positively charged. What were they? Crookes conjectured a 'fourth state of matter': but the true answer was just 'particles of electricity'—electrons. In 1897 J. J. Thomson compared the deflection of these particles by electrical and magnetic forces, and so was able to calculate the mass of the particles and to measure their charge: it turned out that they had about 1/2,000 of the mass of a hydrogen atom. They could be obtained from every kind of matter, so J. J. Thomson concluded that they must be contained in every atom. They had a negative charge, but the atom had no charge, so there had to be something with a positive charge in the atom, but he could not yet discover what it was.

At just about the same time came the discovery of radioactivity. In 1896 Becquerel found that compounds of uranium gave out penetrating rays. Two years later Pierre and Marie Curie isolated radium salts from uranium minerals, such as pitchblende, and this new and intensely active matter astonished the world. But it was more astonished when, about 1902, it was proved, contrary to all nineteenth-century ideas, that radioactive elements were not immutable, but were transmuting themselves, changing into other elements which in their turn changed into others: between 1902 and 1912 the complicated genealogies of the successive changes of these elements were worked out. In the course of this work a startling conclusion was reached by F. W. Soddy, that there could be several atoms differing in weight but having exactly the same properties: these were called *isotopes*.

The radio-elements gave out more than one kind of ray. There were alpha-particles, which were the nuclei of helium atoms moving at a great pace: these were positively charged, heavy, and arrested by a few centimetres of air: there were beta-rays, which were electrons, negatively charged, light but very fast moving and travelling rather farther through air; and there were gamma-rays, which were not particles but short wave radiation like X-rays, and with even greater powers of penetrating matter. How was it proved that there were three kinds of rays? A simple way of showing this is to apply a magnetic field. The alpha-rays are deflected one way, being positive, the beta-rays the other way, being negative, and the gamma-rays, like light, are not deflected at all. These rays did not come in a continuous stream, but as each atom exploded it ejected the appropriate ray or particle. This was first shown by the aid of the spinthariscope invented by Sir William Crookes. A speck of radium was mounted near to a fluorescent screen, which could be viewed through a lens. Eyes sensitized by a few minutes of darkness could then detect bright sparks or flashes on the screen, each due to a single particle striking the fluorescent material.

The discoverers of radioactivity were astonished at the fact that radioactive substances were giving out energy and that a given weight of, let us say, radium would give out far more energy than the same weight of any fuel or explosive. This energy could not be thought to come from nowhere—so it appeared that the atom was not merely a particle of matter, but a source of energy hitherto unthought of. The possibility of tapping this energy occurred to many but several decades were to elapse before any way of eliciting this energy at will could even be visualized as a possibility.

Radioactivity showed us that atoms were structures and also sources of energy; but we were soon to learn their size, weight and number. About 1910 several physicists turned their attention to the problem of counting the atoms or molecules in a given amount of matter; half a dozen different methods of doing this were invented and agreed in estimating the number of molecules in a litre (about a quart) of gas as about 4×10^{22}. Here is one way in which it can be done. Take an extremely minute but weighed speck of radium and hold it near a fluorescent screen, as in the spinthariscope. In darkness you can just see sparks, each being the result of the crash of a helium nucleus (alpha-particle), ejected from a radium atom, hitting the screen. The flashes are counted and simple arithmetic tells us how many of these helium nuclei are given out by a gram of radium in a year. Then some radium—as much as can be begged or borrowed—is sealed up in an evacuated tube. There it remains a year: the helium is then taken out and measured. Thus we know the number of atoms in this volume of helium and can calculate very easily the number of atoms in a litre of it.

Another way of arriving at the same result was, by the aid of a good microscope, to watch minute but visible particles being pushed about by the jostling of molecules in a liquid. This effect—the Brownian movement—has been known for more than a century. Einstein saw that from the distance a particle drifted in a given time he could calculate the number of particles in a litre of gas. The answer agreed reasonably with some seven others, arrived at in quite different ways. Once we knew how many atoms there were in a known weight of matter we could calculate very easily how heavy each atom was, and at least give a good guess at its size.

Now to return to the structure of the atom. J. J. Thomson thought of a cloud of electrons floating in an atmosphere of positive electricity, but in 1911 Ernest Rutherford investigated the atom by shooting it with the high-speed alpha-particles from radium. He took a thin piece of metal (which, of course, consisted entirely of atoms) and let a narrow beam of the alpha-particles pass through it and register themselves on a photographic plate. The result showed that most of the particles were only very slightly deflected from their course, but that a very few were turned through a large angle or even turned back on their track. Since an alpha-particle is about 8,000 times as heavy as an electron it could not be much deflected by one: so clearly there was in the atoms of the metal foil a body which was heavy compared to an alpha-particle (because it deflected it through a large angle) and very small compared to an atom (because nearly all of the alpha-particles missed it). This body Rutherford called the nucleus of the atom. So now in 1911 the atom was thought of, quite rightly, as a cloud of electrons and a minute heavy positive nucleus. The simplest nucleus, that of hydrogen, Rutherford took to be a single positive particle to which he gave the name of proton.

Now we will take up another story, the revelation of the structure of crystals and later of molecules by X-rays. The generation of X-rays dated from 1895, but for many years no one could be sure that they were waves only differing from those of light in having a very much shorter wave-length. Then in 1911 came the brilliant discovery of von Laue. The geometrical forms of crystals were known to be consistent with the notion that they were made up of identical units (atoms or molecules probably) arranged in regular repeating patterns. Von Laue thought that the atom layers in a crystal, a million or ten million to the millimetre, might affect X-rays in the same way as a fine grating of say five hundred lines to the inch would affect light. He sent off a couple of his younger colleagues to try the experiment, which at once showed that a crystal would split up a beam of X-rays into a regular pattern of spots. This proved that X-rays were radiation. Very soon after, Sir William Bragg and his son Sir Lawrence Bragg showed how to discover the positions of the layers of atoms in a crystal by reflecting X-rays from them and finding the angles at which the strongest reflections appeared. This opened up a quite new field. Gradually the Braggs perfected their method and after about ten years, they and others who had followed them had succeeded in working out not only the characteristic patterns in which the atoms or molecules that made up crystals were arranged, but also how the different atoms out of which the crystal was built were placed relatively to each other. Thus it was shown that a crystal of common salt is not a pattern of sodium chloride molecules, but of sodium and chlorine ions, each equidistant from six others. It was shown that in diamond, the hardest of all bodies, all the atoms were chemically combined together and that the whole crystal was a giant molecule. The Braggs and their followers not only showed the world of solid bodies to be a world of wonderful regular patterns, but were able to work out the structures of some molecules for which the chemists had never been able to find formulae. Later on, beams of electrons were used instead of X-rays and then not only the crystal but the molecule itself could be mapped. When, in the last war, the formula of penicillin was urgently needed, to help the chemists in their efforts to sythesize the drug, this technique was used on it. Fig. 115 shows the map they made and the actual formula.

Fig. 113. One of Sir William Bragg's earliest models of crystal structure, that of zinc blende (zinc sulphide).

This work threw new light on the atom also. Bragg had showed that by

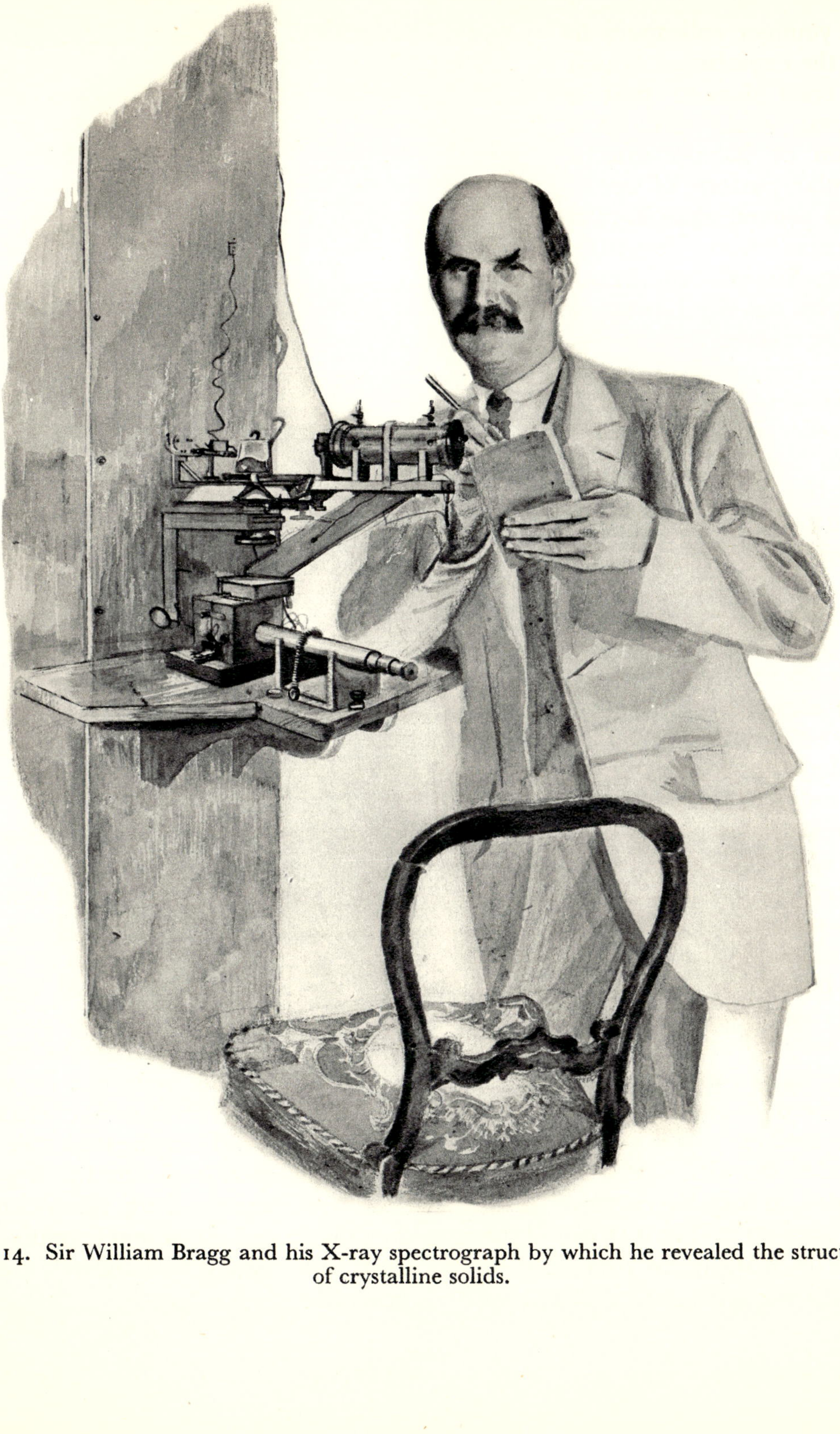

Fig. 114. Sir William Bragg and his X-ray spectrograph by which he revealed the structure of crystalline solids.

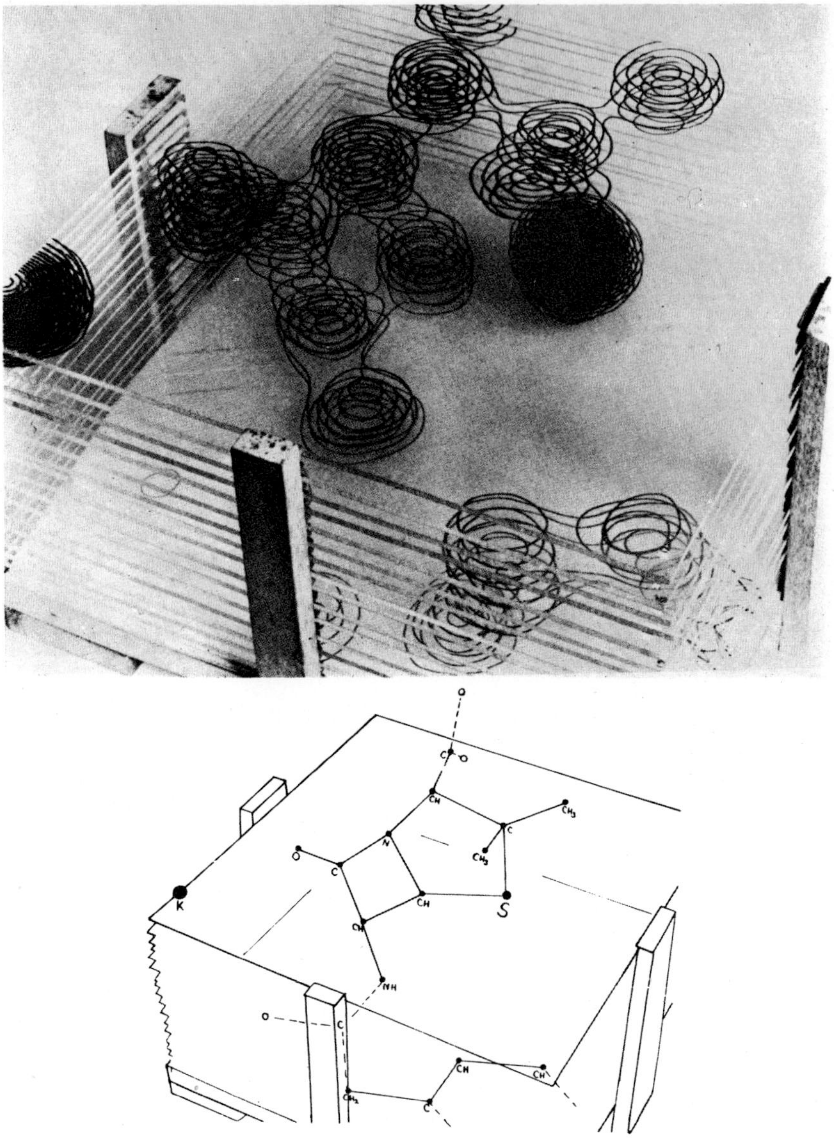

Fig. 115. The distribution of the electrons in the penicillin molecule mapped out in a solid model built up from Perspex sheets. *Below:* The formula as deduced by this and other evidence.

reflecting a beam of X-rays from a crystal it was possible to photograph the spectrum of the rays; that is, to measure their wave-lengths. In 1915 Moseley measured the wave-lengths of the X-rays given out by different elements when used as targets in an X-ray tube. The results led him to an important hypothesis about the numbers of electrons and protons in the atom. If all the elements were

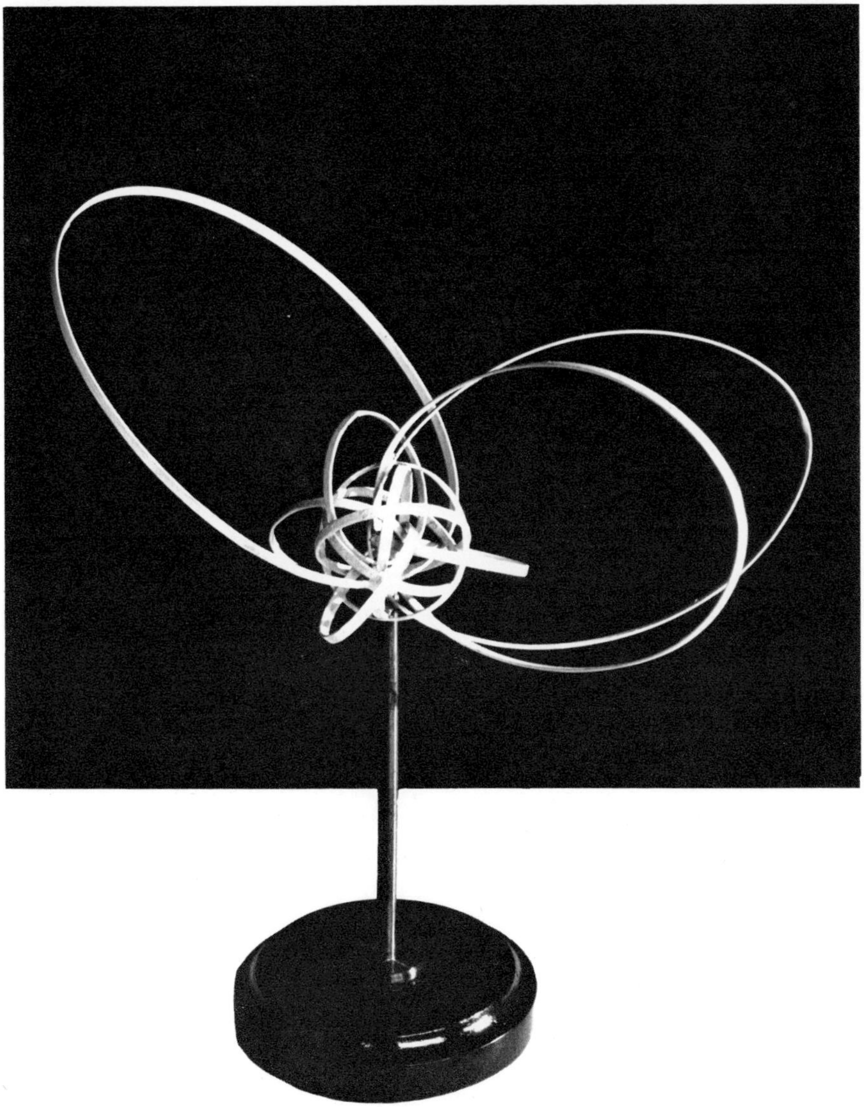

Fig. 116. Model of the aluminium atom illustrating the orbits of electrons as proposed by Niels Bohr.

set out in order of Mendeleyev's table and numbered through, 1, 2, 3, etc., from the beginning, we call the number of each element its *atomic number*. Now every atom, according to Moseley, had as many electrons in its cloud as its atomic number, and as many protons in its nucleus as its atomic weight; the nucleus also contained enough electrons to make the whole atom electrically neutral.

This seemed to follow quite sensibly, but why were the negative electrons not attracted on to or into the positive nucleus? Was it for the same reason that the earth does not fall into the sun, namely, because it is revolving around it? But this explanation would not hold for the electrons, because a revolving electric charge would generate electromagnetic waves, lose its energy and drop into the nucleus. The answer to the problem was Planck's quantum theory. Niels Bohr, from 1913, proposed the theory that the electrons rotating round the nucleus could only possess certain fixed quantities of energy and never less than a certain minimum. So an electron could rotate in certain fixed orbits and no others: when it jumped from one orbit to another it gave out or took in energy, but while it stayed in an orbit it did neither. Now this theory would not have carried much conviction if it had not led to new discoveries, but in fact it explained the very odd fact that glowing atoms (e.g. in a sodium flame or a neon tube) gave out light of only a few wave-lengths (each represented by a spectral line) instead of *all* wave-lengths, as a piece of white-hot iron does. Each line represented one possible orbit-jump and it became possible to explain the reason why the spectra of certain elements contained certain lines. Scientists were amazed. No one had thought that spectra would be explained within fifty years and here were Bohr and his followers doing it.

Bohr's atom, then, had the same nucleus as Moseley's, but the cloud of electrons round it was arranged into groups according to their orbits. From the nucleus outwards the electrons were grouped as follows: first 2 electrons in 1-quantum orbits, then 8 electrons in 2-quantum orbits, then 18 in 3-quantum orbits, and so forth. Soon it was found that, while this plan was correct, there were other groupings within the main groupings, and finally that each electron required four quantum numbers to describe it, one for the orbit, one for the movement of the orbit, one for the electron's magnetic properties and one for its spin, clockwise or anti-clockwise. Pauli put forward the simple rule that no two electrons in an atom could have all four quantum numbers the same, and from this the size of all the groups could be calculated. This work explained Mendeleyev's table of the elements. The groups and periods of this table corresponded to particular electron groupings, and Pauli's principle showed us why there could be just so many patterns. Moseley's work enabled us to give reasons why there were just so many elements lighter than uranium and no more; though the existence of elements heavier than uranium remained unguessed. We had made sense of the apparent chaos of elements.

Now it was the turn of the chemists, who set themselves to answer the question, 'What sticks atoms together to make the molecules of compounds?' The first answer was, 'Atom A gives an electron to atom B. A becomes electrically positive because it has lost an electron, B becomes electrically negative because it has gained one; so they attract each other, like rubbed sealing wax and rubbed glass, and so therefore stick together.' Alternatively, two atoms might be joined by the sharing of an electron. All this threw floods of light on the different kinds of chemical compounds, and the working out of it is still going on.

Let us stop for a minute at 1930 and see what the twentieth century had already done. It had counted, measured and weighed atoms, showed how they were made out of electrons and a positive nucleus, explained why and how they combined to make molecules, mapped out some of the molecules, explained what crystals are and how the atoms and molecules pack together to make them; it had proved that some elements transmuted themselves into others and had showed that a vast store of energy was locked up inside them: it had explained why spectra were made up of lines and in some cases why the lines are where they are.

The electron, discovered in 1897, is of interest not only as one of the three particles of which the material world is made up, for it has also become a tool by means of which the scientist operates a vast number of instruments, some of them of the first importance for everyday life. The electron, being extremely light, is very easily and rapidly moved by electrical and magnetic forces. Consequently a stream of free electrons (such as traverses a cathode-ray tube) can, so to speak, act as a machine capable of action at a far higher speed than any made up of ordinary matter. Among such electronic devices are valves, rectifiers, photoelectric cells and cathode-ray tubes. These rendered possible the modern practice of radio, television and radar—how great a part the first two play in ordinary life we all know, while to the last many of us owe that life itself. But beside these means of communication, electronic devices provide a sensitive means of controlling all manner of industrial operations. The modern plant is not controlled by turning valves and pulling levers: it controls itself by electrical observing devices pre-set to the desired limits, operating the controls. The cathode-ray tube is not only of use for television, but is a most versatile means of telling us what is going on in any sort of electrical observing device: its uses are legion.

Finally the science of electronics has made possible an extremely compact and swiftly acting type of automatic computing machine, the most powerful of all aids to the human mind; such machines enable us to investigate many fields from which the scientist was debarred by the sheer length of time necessary for the computations. How much more these machines will be made to do we have yet to learn.

Now for the greatest of all discoveries, the liberation of some part of the energy of the atom. Radioactivity does that, but in its own time. The abundant radio-elements took billions of years to liberate their energy; the elements that liberated it quickly could be obtained only in unweighably minute quantities,

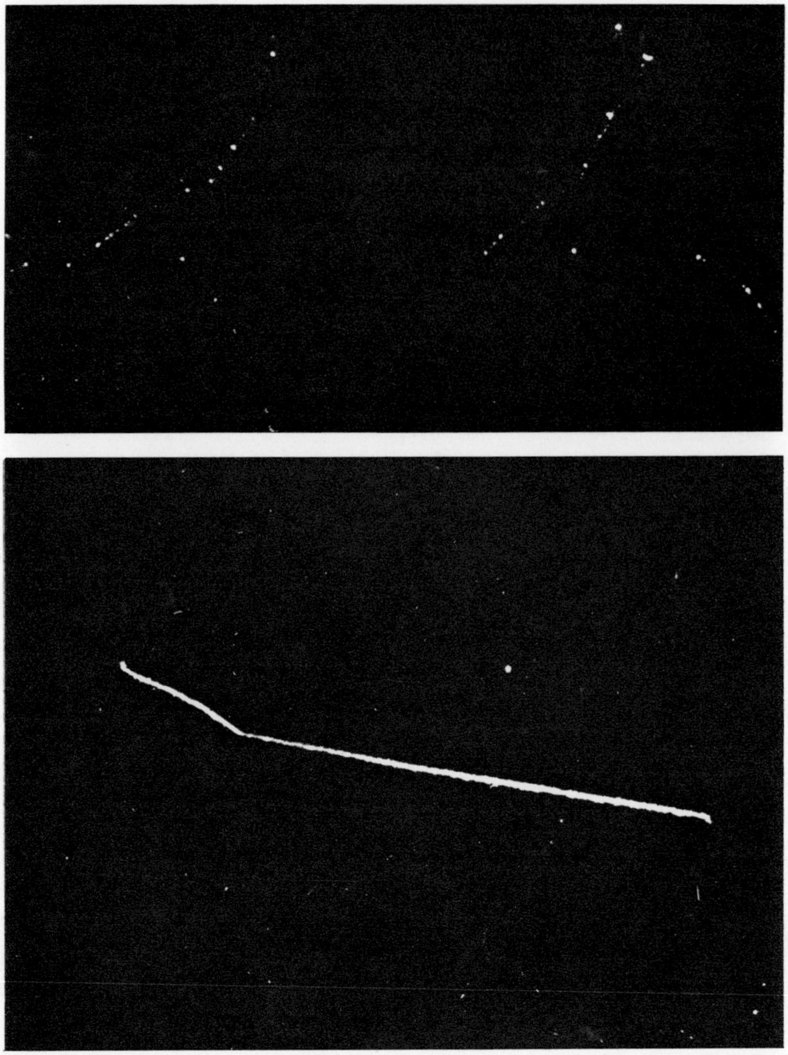

Fig. 117. Photographs of trails made in the cloud-chamber. *Above*, a thin beam of X-rays coming from the left detaches from the air electrons, which make the trails shown (C. T. R. Wilson, 1923): *below*, a neutron, passing vertically upward, splits an atom. The neutron leaves no trail, but the two products show strong trails (N. Feather, 1932).

so the total of energy that could be produced, let us say, in an hour, was very small. But the radio-elements at least provided bullets, so to speak, minute enough but with terrific energy for their size—alpha-particles travelling at 20,000 miles a second or more. These were the first known particles with sufficient energy to penetrate within an atom.

Fig. 118. Professor C. T. R. Wilson operates his invention, the cloud-chamber, perfected by him in 1911–12.

In 1911 C. T. R. Wilson, following up earlier experiments, perfected the cloud-chamber, which enabled us to see and photograph the track of a high-speed particle. In essence it provides air supersaturated with moisture and this is enough to show up the tracks of high-speed particles. As the particle tears through the atoms of the gas it knocks off electrons from them and leaves a trail of electrically charged molecules behind it. Each of these becomes a centre on which a fog-particle—a tiny drop of water—condenses; the result is a line of fog showing the trail of the particle—rather like the visible trails of cloud that invisible aeroplanes leave behind them in the supersaturated upper air. The cloud chamber supersaturates the air by quickly lowering the pressure and immediately takes a photograph. Any atom trails due to radioactive material or other sources are thus registered. Using this apparatus, P. M. S. Blackett in 1919 demonstrated a *forked* atom trail. The thickness and angles of the arms of the fork showed what had happened. A nitrogen atom had swallowed up an alpha-particle, ejected a proton and had been transmuted into oxygen! Work of this kind soon showed that transmutations of one element into another were really quite common, though only on the minutest scale.

In 1932 James Chadwick made an intensely important discovery: he bombarded the element beryllium with alpha-particles and obtained from it a new kind of particle, having the same mass as a proton, but no charge; this he called the neutron. It was soon realized that this was an essential constituent of the atom: that the nucleus was made up in fact of protons and neutrons, not protons and electrons.

The neutron proved to be very useful for bombarding atoms, for, since it had no charge, it was not electrically attracted or repelled and could penetrate the intense field of force round the nucleus. For seven years every effort was made to bring about new transmutations of elements by hitting the atom harder and harder. All manner of apparatus was devised for accelerating particles to the highest velocity and thereby giving them the maximum of energy. By these means it was found possible to transform a small proportion of the atoms of many elements into new unstable atoms, which soon decomposed just like radium and the natural radio-elements.

Then it happened. From 1934 various investigators had bombarded uranium with neutrons but the important step was taken in 1939 when Frisch and Meitner interpreted the results as proving that a uranium atom, struck by a single neutron, could split up into two or more smaller atoms, giving out much energy and several more neutrons. It seemed as if this should make the whole mass blow up, for each of the neutrons liberated could split another atom! But it did not, and only by intense work and the spending of millions of pounds was it found possible to separate from ordinary uranium, a special variety (isotope) called uranium 235, which did explode in this fashion as soon as a sufficient quantity was assembled. If there were less than a critical amount of it, the neutrons escaped without exploding enough uranium atoms to keep up the neutron supply. If there was more than that amount, the uranium atoms were all split up in a minute fraction of a second with liberation of unheard-of quantities of energy. The atomic bomb was with us. The principle behind the bomb is Einstein's mass-energy relationship, $E = mc^2$. When the uranium atom is split, the products weigh very slightly less than the uranium, the balance of its mass being turned into energy. It had long been supposed that there was the same kind of change of mass into energy when hydrogen atoms were built up into heavier atoms in the sun and stars. It seemed that this could be done on earth if the hydrogen could be made hot enough. And it has been done. A uranium bomb provides the heat and the hydrogen has to be of the special kind called heavy hydrogen. The hydrogen bomb can provide far more energy than the uranium bomb, and the world is very worried about it.

It is pretty clear that nothing except goodwill and common sense will prevent these processes being used as weapons and war becoming catastrophic to mankind. But, given those qualities, the reactions that power the bomb may enrich the world. Power-plants energized by uranium and its products are coming into action with results we cannot yet foresee.

In these years, from 1897 to the present day, physics has revealed the structure of matter, and as matter in some form is the subject of every science, there has been scarcely any science that has not wonderfully advanced as a consequence.

Some of the effects upon chemical technique have already been mentioned. The boundary between chemistry and physics has been replaced by a common ground, the study of atoms and molecules with the forces that influence them. The large-scale techniques of chemistry of course remain and are the mainstay of research, and they have been refined by the application of the techniques of physics: outside the atomic and molecular field we see in chemistry a continual and rapid advance, but not a revolution.

On another flank the boundary of chemistry and biology has become the scene of intense activity. While atomic physics is now the centre of the growth and activity of science, biology is creeping up on it and it may well be that in half a century we may find therein the way of modifying not only our surroundings but ourselves. The work of the last fifty years, in biology as elsewhere, has centred upon the discovery of the existence and nature of the minute units that co-operate in the life-process. The truth of Lavoisier's saying that life is a chemical function has come home to us: we now think of the living organism in terms of minute parts at or below the limit of microscopic visibility, each being the scene of elaborate chemical processes of a type very different from those of the laboratory.

Fig. 119. The old way. W. H. Wollaston, about 1805, who, when asked by a friend to show him his laboratory, instructed the footman to bring it in on a tray.

Fig. 120. The new way. The heavy machinery of the modern physicist, as typified by the Birmingham cyclotron.

The study of these units calls for observation of and experiment upon the extremely small. To see or photograph the minute is a matter of physics. The ordinary microscope had come fairly near to perfection by 1900, but in the last few years improvements have begun again. The reflecting microscope with metal mirrors instead of lenses gives a distincter image and allows the use of ultra-violet light: the phase-contrast microscope enables objects too transparent and colourless to be visible to be thrown into sharp contrast with the surrounding medium. An even greater advance has been made by the use of the electron microscope which employs a beam of electrons instead of light and electromagnets instead of lenses. Magnifications have thus increased fifty times, though the difficulty of preparing and handling the materials to be magnified limits the range of that which can be studied. Finally, the X-ray technique developed from the work of Bragg and von Laue enables structures of atomic dimensions, if not to be seen, at least to be mapped. The result of all this is that we know enormously more about the structure of the living being, though what we have still to learn much exceeds what we already know. To see is not enough: we need to experiment. In the last decade micro-manipulators, operating minute needles and hooks, syringes and knives by means of fine-adjusted screws, allow bodies invisible to the eye to be dissected and manipulated.

We are still very ignorant of the nature of the material of which living beings are made. Only very recently has it become possible to assign a chemical formula to even one of the simplest of proteins, of which millions, all different, are to be found in living organisms. Nevertheless we are now able to distinguish many more of the parts of the cell and assign to them their functions. The greatest of these discoveries was that of the mechanism of heredity: the existence of the thousands of hereditary factors strung together to form the chromosomes of the cell, and the way they are passed on from parent to offspring. It is a great thing to know that here is the pattern that made me what I was born, but we have still to discover the mechanism by which this array of structures, built up of unknown molecules, determines the qualities of the being that possesses them. These chromosomes are only one of the many structures observed in the cell which was, a hundred years ago, regarded as a mass of structureless jelly.

The techniques of chemistry applied to living beings have borne wonderful fruit. In the earlier years of the century it was enough to use the ever more refined methods of the organic chemist to discover new substances that played an essential part in the life process. The vitamins, relatively simple substances but essential in animal nutrition, the hormones and endocrines by which one organ of the body may influence the functions of others, were discovered by the biochemists.

New techniques, electron-diffraction, chromatography, the use of radioactive tracer-atoms, have accelerated the discovery of what living matter is and what is contained in it. Perhaps the most important of advances is towards the understanding of the general nature of the chemistry of the cell. We see that it is based on enzymes, complicated substances that can somehow guide the chemical processes in a particular direction. We see that the cell's chemistry differs from that of our test-tubes and beakers by its wonderful control: we begin to visualize the delicate balance of factors by which energy is *eased* out of foodstuff and oxygen. It is all far more complex than we thought: but it is a great advance even to see the outlines of its general principles, far as we may be from being able to specify details. The foundations of biology are perhaps at the same stage as was atomic physics about 1905. The problems that it presents are probably much more complex and difficult to solve, and, if I may hazard a guess, those who are children now may hope in their old age to have a fair understanding of what we describe as life.

Our knowledge of the universe has increased beyond all reckoning. At the beginning of the century the distances of only a few of the nearer stars could be measured, but today we can estimate the distances even of the most remote objects. Our knowledge of the structure and energy relationships of atoms has enabled us to understand whence the sun and stars generate their heat. Finally the spectra of the most distant objects show us that the universe is expanding, moving ever outwards from us, and we are trying to conjecture what this means.

Is it not a most strange and wonderful thing that the intellects of us little creatures crawling on the surface of a minor planet of a second-rate star in a very ordinary galaxy, can reach out with our senses and minds to stars billions of miles away, to times thousands of millions of years ago, to the particles that are the foundations of things. Are we not greater wonders than the atoms and the stars?

APPENDIX I

NOTES UPON ILLUSTRATIONS

THE FOLLOWING NOTES ARE INTENDED to inform the reader how far the drawings specially made for this book are based upon records and how far upon inference and imagination. Having regard to the occasion and character of the book, I have not thought it necessary in every case to indicate the source of the portraits used.

Figure 2 Based on the drawings and specimens in the Science Museum. The manner in which the hour-watchers made their observations is conjectural.

Figure 4 The water-clock and sundial are based on actual specimens.

Figure 5 The alarm clock is drawn from a fifteenth-century specimen, by courtesy of its owner Capt. Antonio Simoni.

Figure 6 The scene is based on the pulpit at St. Andrews, traditionally connected with John Knox. This pulpit is probably, in part at least, later than the time of Knox, and we have taken the liberty of making some minor alterations.

Figure 7 Based on a well-known relief.

Figure 8 The source is various tomb-paintings, some from Rekhmiré. (Fifteenth century B.C.)

Figure 9 Thales is here supposed to have thought of his theory, that water is the source of all things, during his travels in Egypt and particularly in the Nile delta, where the water seems to produce earth, while sea and sky seem to melt one into another.

Figure 13 The likeness of Simon Stevin is taken from a contemporary portrait by an unknown artist. The remainder of the picture is an interpretation of his description (p. 20) of the experiment against Aristotle.

Figure 14 Aristotle, when newly married, observed the life of sea-creatures on the shores of Mytilene. Various busts of Aristotle, none of any great authenticity, have provided the artist with his model. The portrait of his young wife is imaginary.

Figure 20 No scaph has survived from Græco-Roman times. The figure shows the construction but we have no authority for the decoration.

Figure 22 Pythagoras is said to have discovered the proportion of length to pitch in strings. Since pitch depends on mass and tension as well as length, the former must have been kept constant throughout the experiment, and this could best be ensured by using the same string. I suppose, then, that Pythagoras measured the distances from the bridge to the frets on some stringed instrument, like a lute or guitar. The Greeks of the time of Pythagoras are not known to have had such instruments, but they are figured in Egyptian drawings. Pythagoras went to Egypt—hence our scene. I am indebted to the British Museum for information and for a photograph showing the instrument, but it must not be supposed that our reconstruction is sponsored by that institution. The portrait of Pythagoras is of course imaginary. He is measuring the instrument with a cubit-scale, the drawing of which is based on an actual specimen.

Figure 24 Nothing is known of Philo's appearance and his date is conjectural. The apparatus is based on his descriptions and an illustration in MS Digby 40 in the Bodleian library.

Figure 25 The diagram is copied from an illustration deriving from a MS, and doubtless reproduces Hero's original illustration. The picture attempts to show the effect upon the devout.

Figure 26 The portraits of Hero and his friends are imaginary. The picture of the turbine is based on drawings deriving from the MSS.

Figure 27 St. Augustine is shown as a young man from North Africa, not as the bearded bishop of Christian iconography. No authentic portrait exists.

Figure 29 There is no authentic portrait of Mary the Jewess. The scene is Alexandria. The apparatus is based on the drawings in the Greek Alchemical MSS. These are reproduced by Berthelot. (Collection des Alchimistes Grecs. vol. 1. Introduction. 1887, pp. 127–73.)

Figure 34 There is no authentic portrait of Roger Bacon. The incident portrayed is merely inferred from his text, as quoted on p. 48.

Figure 35 Based on numerous contemporary representations.

Figure 36 Based on a scene portrayed upon a Greek vase. (See Fig. 5 Singer. C. Greek Biology and Greek Medicine. Oxford, 1922.)

Figure 38 Based on Galileo's description and numerous portraits. There has been no attempt to represent any particular room.

Figure 42 Based on Galileo's description and figure in his *Dialogues concerning two new Sciences.*

Figure 43 Based on Torricelli's own illustration of the apparatus, and on a portrait of Torricelli. Galileo's picture is hung on the wall to show that he was Torricelli's master.

Figure 45 Galileo is shown as a very young man. The scene is conjectural: the lamp similar to but not identical with that shown at Pisa, but in fact not placed there until after Galileo's death.

Figure 47 Galileo is shown making observations of the moon during the visit that he paid to Venice in 1609 in order to show his telescope to the Doge and Senate. There is no record that he observed the skies during these few days, but he would scarcely have missed the opportunity of a clear night.

Figure 51 No portrait of Hooke survives but he was noted for his ill looks. Samuel Pepys speaks of 'Mr. Hooke who is the most, and promises the least of any man in the world that I ever saw'. The microscope is taken from Hooke's *Micrographia*, 1664.

Figure 53 The likeness of Leeuwenhoek is taken from the well-known portrait painted and engraved by J. Verkolje.

Figure 55 The table of instruments is taken from the *De Fabrica* of Vesalius in the belief that there would have been but little change in the eighty years or so between his time and that of Harvey.

Figure 57 The room is not intended to represent any existing apartment. The details of the positioning and support of the prism do not appear in Newton's works, but the latter was probably much closer to the shutter than we have represented it.

Figure 59 Based on the actual instrument, in the possession of the Royal Society. The bench symbolizes the fact that Newton was skilled with his hands.

Figure 62 The portraits are authentic: the room imaginary.

Figure 65 The likeness of von Guericke is based on a contemporary portrait. The elegant figure is intended to represent M. Monconys who visited von Guericke and saw his electric machine. In the background, activity towards the experiment of the Magdeburg hemispheres.

Figure 67 Based upon the portraits of Granvil Wheler at Otterden Place, and on the gallery of that house, by kind permission of the present owner Granville Wheler Esq. No portrait of Stephen Gray survives: he is shown as a Brother of the Charterhouse. The apparatus is based on the description in *Phil. Trans.* XXXVII. 1731, pp. 19–44. See also Appendix II.

Figure 68 Based on du Fay's description of the experiment, summarized by Joseph Priestley. *Hist. and Pres. State of Electricity.*

Figure 69 Based on the Abbé Nollet's numerous illustrations in his *Essai sur l'electricité des corps, 1746,* and the artist's imagination.

Figure 73 The figure of Boyle is based on numerous portraits. The air-pump is figured in *New Experiments Physico-Mechanical touching the Spring of Air,* 1660. For Hooke, see note on Fig. 51.

Figure 77 The apparatus is based on the illustrations in Cavendish's paper in *Phil. Trans.* for 1766. The portrait of Cavendish is an attempt to infer his appearance at a date much earlier than the only surviving portrait.

Figure 78 Based on a portrait and illustrations of apparatus shown in Scheele's works.

Figure 80 Based on portraits of Lavoisier and his wife, her drawings of their laboratory, and apparatus shown in his works, especially the *Traité Élémentaire de Chimie.*

Figure 82 An attempt to show Dalton teaching private pupils, one of whom was Joule, at the Manchester Literary and Philosophical Society.

Figure 83 Modified from a picture in the Walker Art Gallery, Liverpool.

Figure 84 The cannon and arrangements for demonstrating the heating effect are taken from Rumford's paper in *Phil. Trans.* for 1799: the horse-mill is drawn from a model in the Science Museum of one formerly used for the boring of cannon at Woolwich Arsenal, with inessential modifications intended to display the working of the mill more clearly.

Figure 87 Based upon a portrait of Galvani and the descriptions and figures in his collected works, Bologna, 1841. No attempt has been made to draw the actual veranda, and I do not know whether it exists.

Figure 88 Based upon a portrait of Volta and illustrations in his collected works, and *Phil. Trans.* for 1800.

Figure 89 Based upon Davy's description of the experiment and on early pictures of the lecture-room of the Royal Institution. The placing and manner of arrangement of Davy's large battery is conjectural.

Figure 91 Based on a portrait of Oersted and his descriptions of the experiment (Thomson's *Annals of Philosophy*, Oct. 1820). Oersted, in fact, used a box-compass, which has been preserved, but a needle is here shown to demonstrate the effect more clearly. There is no indication in this account of the manner in which the wire was supported over the compass. The background is imaginary.

Figure 95 Based on the apparatus at the Royal Institution, and pictures of Faraday and his laboratory.

Figure 97 Based on Lord Kelvin's description (*Nature*, vol. XXVI, p. 618, 1882) and the actual terrain. Joule was still a young man: his appearance is inferred from a later portrait.

Figure 99 Based on the illustration in *Phil. Trans.* for 1800 and portraits of John and Caroline Herschel.

Figure 101 Based upon a photograph of Hertz and pictures of his apparatus, which is preserved at Munich in the Deutsches Museum. The battery and coil have not been preserved, but those represented are of the type then in use.

Figure 103 Drawn from the actual tubes and coil now in the Science Museum, from a contemporary battery, and from several photographs of Crookes.

Figure 105 From a portrait of Bunsen and an early illustration of his spectroscope.

Figure 108 The figure of Dr. Paris was copied from his portrait by kind permission of its owners, The Royal College of Physicians. The apparatus for liquefying chlorine is described but not figured in Faraday's paper.

Figure 109 This is based upon several illustrations in *The Collected Papers of Sir James Dewar*, ed. Lady Dewar, Cambridge, 1927. While the apparatus is authentic, it probably was not all assembled in the same place.

Figure 111 Apart from the figures of Pasteur and the apparatus on the table, this is a work of the imagination.

Figure 112 No such scene is known to have taken place, but Darwin must have observed the chimpanzees which throughout his life were in the Zoological Society's collections. It symbolizes, at least, the link that Darwin propounded between the human and animal species.

Figure 114 Based upon a photograph of Sir William Bragg, with his X-ray spectrograph, kindly furnished by Sir Lawrence Bragg.

Figure 118 From a photograph of Prof. C. T. R. Wilson, the illustrations and descriptions in his papers in *Phil. Trans.* 1911, p. 285, supplemented by further information kindly furnished by Prof. Wilson himself.

Figure 119 Based on the well-known portrait of Wollaston and the anecdote in J. A. Paris's *Life of Sir Humphry Davy* and Wilson's *Religio Chemici*.

Figure 120 Based on the Birmingham Cyclotron. The comparison is not really a fair one, for the successor of Wollaston might be using apparatus even more minute; but perhaps it will serve to underline the new tendency to employ great forces upon small particles of matter.

APPENDIX II

STEPHEN GRAY

STEPHEN GRAY HAS RECEIVED LITTLE notice in the histories of science. It has therefore seemed to me worth while to transcribe a curious poem, the circumstances of the production of which appear in Boswell's *Life of Johnson.* (Oxford edition, vol. II, 351.)

> 'He (Johnson) published nothing this year (1766) in his own name; but he furnished the Preface and several of the pieces which compose a volume of *Miscellanies* by Mrs. Anna Williams, the blind lady who had an asylum in his house. There is in this collection a poem, "On the death of Stephen Gray, the Electrician", which, on reading it, appeared to me to be undoubtedly Johnson's. I asked Mrs. Williams whether it was not his. "Sir (said she, with some warmth), I wrote that poem before I had the honour of Dr. Johnson's acquaintance." I, however, was so much impressed with my first notion, that I mentioned it to Johnson, repeating, at the same time, what Mrs. Williams had said. His answer was, "It is true, Sir, but she has not told you that I wrote it all over again, except two lines."'

On the DEATH of STEPHEN GREY, F.R.S.
The Author of
*The Present Doctrine of Electricity**

Long has thou born the burthen of the day,
Thy talk is ended, venerable GREY!
No more shall Art thy dext'rous hand require
To break the sleep of elemental fire;
To rouse the pow'rs that actuate Nature's frame,
The momentaneous shock, th' electrick flame,
The flame which first, weak pupil of thy lore,
I saw, condemn'd alas! to see no more.

Now, hoary Sage, pursue thy happy flight,
With swifter motion haste to purer light,
Where BACON waits with NEWTON and with BOYLE
To hail thy genius, and applaud thy toil;
Where intuition breaks through time and space,
And mocks experiment's successive race;
Sees tardy Science toil at Nature's laws,
And wonders how th' effect obscures the cause.

* The Publisher of this Miscellany, as she was assisting Mr. Grey in his experiments, was the first that observed and noticed the emission of the electrical spark from a human body.

Yet not to deep research or happy guess
Is ow'd the life of hope, the death of peace.
Unblest the man whom philosophick rage
Shall tempt to lose the Christian in the Sage;
Not Art but Goodness pour'd the sacred ray
That cheer'd the parting hour of humble Grey.

INDEX

Acceleration, 19, 54, 56
Action at a distance, 40
Air, 16, 35, 94–102
Air, liquid, 142–3
Air, weight of, 57
Air-pump, 60, 61, 95
Albertus Magnus, 22
Alchemists, 42, 43, 45, 102
Alcohol, 45
Alembic, 43
Alexandria, 31, 32
Alpha-rays, 152
Amber, 40
Ampère, 118
Anatomy, 51, 52, 53, 68, 72
Arago, 118
Arc-light, 115, 117
Archimedes, 32
Aristarchus of Samos, 30, 31, 52
Aristotle, 16, 20–3, 28, 41, 53, 102
Assaying, 45
Astrolabe, 23, 24
Astronomical observation, 22, 62, 134
Astronomy, 3, 32, 62–3, 134–9, 163
Atom, 20
Atom, structure of, 151–8
Atomic bomb, 159–63
Atomic number, 156
Attraction, 40, 82
Augustine, 39, 40

Bacon, Francis, 68
Bacon, Roger, 46, 47
Bacteria, 145–6
Balances, 10
Battery, electric, 113–14
Becquerel, Henri, 151
Beta-rays, 152
Biochemistry, 163–6
Biology, 22, 123, 163
Black, Joseph, 98
Blackett, P. M. S., 162
Bohr, Niels, 157
Boyle, Robert, 14, 61, 84, 95
Bragg, Sir Lawrence, 153–4
Bragg, Sir William, 153–4
Brahe, Tycho, 26
Bunsen, Robert, 134, 137

Callippus, 28
Caloric, 108
Capillaries, 72
Cathode rays, 134, 151
Cathode-ray tube, 134, 159
Cavendish, Henry, 93, 98–9
Cell, 144, 163
Chadwick, James, 162
Chemical compounds, 105, 158
Chemistry, 42–5, 94–102, 142–3
Chinese, 14, 40
Chromosomes, 166
Circulation of the blood, 72
Civilization, 1
Clock, 6, 8
Cloud-chamber, 160, 161
Combustion, 95–102
Compass, 40
Computing machine, 159
Conductors, 84
Conservation of energy, 128
Conservation of mass and energy, 150
Conservation of matter, 128
Constellations, 3, 5, 10, 52
Copernicus, 29, 30, 66, 67
Coulomb, 93
Craft, 3
Crookes, William, 134, 135, 151, 152
Crystals, structure of, 154–5
Curie, Marie, 151
Cuvier, Baron, 121

Dalton, John 105–8
Darwin, Charles, 146
Darwin, Erasmus, 122

Davy, Sir Humphry, 108, 115, 117
Descartes, René, 120
Dewar, James, 142
Dissection, 51
Distillation, 40
Drugs, 142
Du Fay, 88, 89
Dynamo, 126

Earth, 16
Earth, age of, 120–3, 146–8
Earth, circumference of, 31
Edison, T. A., 126
Egypt, 1, 3–13
Einstein, Albert, 150, 153, 162
Electric battery, 113–14
Electric machine, 84
Electric shock, 88, 90
Electricity, 84–93, 113–19
Electricity and magnetism, 115–17, 124–6
Electrolysis, 113
Electromagnetic machines, 126
Electromagnetic theory of light, 130, 132
Electromagnets, 118
Electronic devices, 159
Electrons, 134, 151, 159
Electroscope, 84
Element, 16ff, 102, 105
Endocrines, 166
Energy, 124, 126, 128
Epicycle, 29
Eratosthenes, 31
Ether, 130, 150
Eudoxus, 28
Evolution, 146–8

Falling bodies, 20, 21, 53–4
Faraday, Michael, 118, 124, 126, 130, 132, 140
Ferdinand, Grand Duke of Tuscany, 62
Fire, 16
Fixed air, 98
Formula, chemical, 105, 106
Fossils, 120, 121
Franklin, Benjamin, 91
Frisch, 162

Galileo, 20, 53–67, 151
Galvani, Luigi, 113, 114
Gamma-rays, 152
Gases, 36, 95–102, 140–3
Gases, liquefaction of, 140–3
Gases, permanent, 140
Generation, spontaneous, 145
Genesis, 120
Geometry, 12, 27
Germs, 145–6
Gilbert, William, 81, 82
Glasses, 12
Glazes, 12
Gray, Stephen, 84, 85–7, 172
Greek fire, 47
Greek Science, 1, 14–51
Guericke, Otto von, 60, 61, 95
Gunpowder, 46, 47, 95

Hales, Stephen, 97–8
Harvey, William, 72–3
Hauksbee, Francis, 84, 88
Heart, 72
Heat, 108, 126, 128, 140
Helmholtz, von, 128
Helmont, J. B. van, 87
Henry, Joseph, 118
Hereditary factors, 166
Hero of Alexandria, 36, 37, 38
Herschel, Sir John, 136, 138–9
Herschel, Sir William, 128, 131, 136
Hertz, Heinrich, 132, 133
Hertzian waves, 132, 133
Hipparchus, 29
Hooke, Robert, 68, 95, 120
Hormones, 166
Hour-watchers, 3, 4
Hutton, James, 122
Huygens, Christian, 61
Hydrogen, 98

Incendiary mixtures, 46
Infectious disease, 146
Insects, 78, 80
Instrument makers, 80, 81
Invisible light, 128, 130, 131
Isotopes, 152

Johnson, Dr., 172
Joule, J. P., 126, 129
Jupiter, 65

Kant, Immanuel, 122
Kepler, Johann, 19, 29, 67
Kirchhoff, 134
Knox, John, 9
Koch, Robert, 146

La Place, P. S. de, 122–3
Laboratory, 47, 78, 80
Lamarck, 122
Laue, von, 153
Lavoisier, Antoine Laurent, 100, 102, 103, 123, 128
Leeuwenhoek, Antony van, 70, 71, 74
Leibniz, Gottfried Wilhelm, 120
Leonardo da Vinci, 52, 56, 72
Leyden jar, 88
Life, 123, 144
Light, 74–8, 41–3, 130
Lightning, 91
Liquid air, 142–3
Lister, Lord, 146
Loadstone, 39, 40, 81
Lucretius, 40

Magdeburg hemispheres, 60, 61
Magnet, 40, 81–3, 118
Magnetism, 81–3, 115–17
Malpighi, 74
Man, pedigree of, 146
Maria the Jewess, 41
Mass-energy relationship, 150
Maxwell, James Clerk, 132
Mayer, Robert, 128
Mayow, Robert, 95–6
Medicine, 13, 48
Meitner, 162
Mendeleyev, 144, 156
Mesopotamia, 1, 3–13
Metallurgy, 12
Micro-manipulators, 163
Microscope, 68–71, 144, 163
Microscope, electron, 163
Microscope, phase-contrast, 163
Microscope, reflecting, 163
Molecules, 105
Moon, 62
Moseley, 156
Motion, 18ff, 53–7

Nature, regularity of, 11
Natural selection, 146
Nebula in Orion, 138–9
Neutron, 162
Newton, Isaac, 68, 74–8, 84, 111, 151
Nicholson and Carlisle, 113
Nile, 3, 15
Nitric acid, 45
Nitrogen, 100
Nollet, 88, 90
Norman, Robert, 81
Nuclear fission, 159ff
Nucleus of the atom, 153

Oersted, John Christian, 115, 117
Organic chemistry, 142
Oxygen, 95, 100

Paris, J. A., 141
Pasteur, Louis, 144–7
Pauli, 158
Pendulum, 61, 63
Penicillin, 154, 156
Perfect gas, 140
Pharmacy, mediaeval, 49
Philo of Byzantium, 35
Phlogiston, 97, 99, 100
Photography, 136, 138
Physics, 163
Physiology, 51
Pile, Voltaic, 113–14
Planck, Max, 150, 158
Planets, 26–9, 65–8
Plato, 17
Polarization, 112, 130
Pollaiuolo, Antonio, 52
Potassium, 113–14
Priestley, Joseph, 100
Projectiles, 56, 57
Proton, 153
Protoplasm, 144
Pterodactyl, 121
Ptolemy, 29

Putrefaction, 145–6
Pyramid, 12
Pythagoras, 32, 33

Qualities, four, 16
Quantum theory, 150, 158

Radiation, 128
Radioactivity, 151–3
Radium, 151–3
Redi, 145
Regenerative cooling, 142
Regularity of Nature, 11
Relativity, 150
Respiration, 95–102
Röntgen, 134
Royal Institution, 124
Royal Society, 81, 83
Rumford, 108
Rutherford, Daniel, 100
Rutherford, Ernest, 153

Sand-glass, 6, 8, 9
Santorio, 62
Scaph, 30
Scheele, Karl Wilhelm, 100, 101
Scheuchzer, 121
Schleiden, 144
Schwann, 144
Science, 3, 11, 14, 78, 81, 105, 150
——, periods of, 1
Sepsis, 48, 146
Soddy, F. W., 152
Sodium, 113–14
Sound, 32
Spectroscope, 134, 137
Spectrum, 75–6, 134, 158
Spinthariscope, 152
Spirit, 43
Star, 24, 134
Steam-engine, 38, 109
Stevin, Simon, 20, 53
Still, 41, 44
Sturgeon, William, 118
Sumerians, 3
Sun-spots, 65
Sundials, 6
Surgery, 48, 146
Swammerdam, 78, 80
Synthetic dyes and drugs, 142

Telescope, 62
Telescope, reflecting, 77, 78
Terrella, 81–3
Thales, 15
Theophrastus, 22
Theories, 1, 14
Thermometer, 61–2
Thompson, Benjamin, 108
Thomson, J. J., 151
Thomson, W., Lord Kelvin, 129
Time-telling, 6, 7
Torricelli, 57, 59
Transmutations, 151, 162

Universe, 18
Uranium, 151
Uranus, 128

Vacuum, 19, 58
Vacuum flask, 142
Venice, 64
Venus, 65
Vesalius, Andreas, 52, 114, 116
Virchow, 144
Vitamins, 166
Volta, Alessandro, 113, 116
Voltaic battery, 113–14

Wallace, Alfred Russell, 146
Water, 15, 16, 102
Water-clock, 6
Wave-theory of light, 111, 130
Wedgwood, Josiah, 138
Weights, 10
Wheler, Granvil, 87
Williams, Anna, 172
Wilson, C. T. R., 160, 161
Wohler, 123

X-rays, 134, 153–4

Young, Thomas, 111, 126

Zodiac, 10